MW00427365

Seeing Red:

Redshifts, Cosmology and Academic Science

Halton Arp

Apeiron
Montreal

Published by Apeiron
4405, rue St-Dominique
Montreal, Quebec H2W 2B2 Canada
http://redshift.vif.com

© Halton Arp

First Published 1998

Canadian Cataloguing in Publication Data

Arp, Halton C., 1927-
    Seeing Red: redshifts, cosmology and academic
science

Includes index.
ISBN 0-9683689-0-5

1. Red shift. 2. Quasars. 3. Galaxies. I. Title.

QB857.A764 1998   523.1'1   C98-900497-X

Figures 7-8, 9 and 10 are reproduced and adapted from
Astronomy and Astrophysics Review 5, 239-292 (1994) with
permission of the authors and Springer-Verlag, Berlin-
Heidelberg, 1994.

Front cover:
A picture in X-rays of the Seyfert galaxy Markarian 205.
Compare to optical image of NGC4319/Mark 205 on the cover
of *Quasars, Redshifts and Controversies* (published by
Interstellar Media in 1987). See also Figure 1-7 and Figure 6-15
here.

Back cover:
The ejecting E galaxy Arp 105. See Figure 3-12 in Chapter 3 for
redshifts of objects in the field. Image from picture by Duc and
Mirabel.

Table of Contents

Preface

My purpose in publishing this book is to communicate information which would not otherwise be accessible. About 10 years ago my first book on this subject appeared: *Quasars, Redshifts and Controversies*. That first book had really been written between 1984 and 1985, but it took a seemingly interminable two years to publish because uncountable numbers of publishers turned it down. One university press, that of my *alma mater*, was enthusiastic about it until they gave it to a member of the Astronomy faculty to read. Another, Cambridge University Press, declined to publish it, but once it was published bought a large number of copies at very low cost to sell through their distribution. (At least the distribution was a useful step).

Finally, Donald Goldsmith came to the rescue of what I view as academic freedom of communication and published it under the aegis of his small company, Interstellar Media. I felt enormously grateful to him for enabling the observational material to be presented, regardless of what he or any one else felt about the ultimate outcome of the debate. Of course, I was hoping that once all the evidence was correlated and described in a way not allowed by referees, scientists would turn their instruments and analysis to investigating the many crucial objects which contradicted current theory.

Instead, the book became a list of topics and objects to be avoided at all cost. Most professional astronomers had no intention of reading about things that were contrary to what they *knew* to be correct. Their interest usually reached only as far as using the library copy to see if their name was in the index. But before that disappointment really registered with me, something rather wonderful happened. I started getting letters from scientists in small colleges, in different disciplines, from amateurs, students and lay people. The amateurs in particular amazed and delighted me, because it quickly became clear that they really *looked* at pictures, knew various objects and reasoned for themselves while maintaining a healthy skepticism toward official interpretations. As an example, Canadian physics students brought me from Europe to address their annual convention. I was stunned when they ushered me into a room where a table was piled high with copies of my book to autograph. I realized that these were books they had bought on their own initiative and with their own money. In the end, the book was translated into Italian and Spanish, and I still hear from people all over the world who are interested in how it is all going to turn out. So regardless of the difficulties and frustrations, and no matter what else happens, I feel that book was the most important and rewarding work I have ever undertaken.

Halton Arp, *Seeing Red: Redshifts, Cosmology and Academic Science* (Apeiron, Montreal, 1998)   i

More than 10 years have passed and, in spite of determined opposition, I believe the observational evidence has become overwhelming, and the Big Bang has in reality been toppled. There is now a need to communicate the new observations, the connections between objects and the new insights into the workings of the universe— all the primary obligations of academic science, which has generally tried to suppress or ignore such dissident information. In spite of—or because of—the success of the first book, it is even more necessary now to secure independent and effective publication of these kinds of science books. The present volume is a bigger book with prospects for wider circulation. In consideration of these aspects, with Don Goldsmith's advice and assistance, I feel fortunate that the present publisher, Roy Keys, is presenting this new work, *Seeing Red: Redshifts, Cosmology and Academic Science.*

One useful aspect of the present book is that it illustrates what can develop from one simple assumption, such as the nature of extragalactic redshifts. Both sides in the dispute have complex, rather fully worked out views which they believe to be empirically supported and logically required. Yet one side must be completely and catastrophically wrong. It makes one wonder, perhaps with profit, whether there are other uncertain assumptions on which much of our lives are built, but of which we are innocently overconfident.

The present book is sure to outrage many academic scientists. Many of my professional friends will be greatly pained. Why then do I write it? First, everyone has to tell the truth as they see it, especially about important things. The fact that the majority of professionals are intolerant of even *opinions* which are discordant makes change a necessity. Those friends of mine who also struggle to get the mainstream of astronomy back on track mostly feel that presenting evidence and championing new theories is sufficient to cause change, and that it is improper to criticize an enterprise to which they belong and value highly. I disagree, in that I think if we do not understand why science is failing to self-correct, it will not be possible to fix it.

Briefly, I suppose my view is that science never matured through the "age of enlightenment." When society at long last learned that major decisions were too important to be left in the hands of kings and generals, a more democratic process was evolved. But science always insisted that only those who possessed arcane knowledge were capable of deciding what was true and what was not true in the world of natural phenomena.

Now we have a situation where new facts are judged by whether they fit old theories. If they do not, they are condemned with the judgment:

"There is no way of explaining these observations, so they cannot be true."

That encourages the dissident to come up with an explanation of how it could be true. It disagrees with convention. Then the jaws of the trap spring shut and the theory is labeled:

"....prima facie evidence that the proponent is a crackpot and the evidence is false."

This, then, is the crisis for the reasonable members of the profession. With so many alternative, contradictory theories, many of them fitting the evidence very badly, abandoning the accepted theory is a frightening step into chaos. At this point, I believe we must look for salvation from the non-specialists, amateurs and interdisciplinary thinkers—those who form judgments on the general thrust of the evidence, those who are skeptical about any explanation, particularly official ones, and above all are tolerant of other people's theories. (When the complete answer is not known, in a sense everyone is a crackpot—Gasp!).

The only hope I see is for the more ethical professionals and the more attentive, open-minded non professionals to combine their efforts to form a more democratic science with better judgment, and slowly transform the subject into an enlightened, more useful activity of society. This is the deeper reason I wrote this book and, although it will cause distress, I believe a painfully honest debate is the only exercise capable of galvanizing meaningful change.

If there is any credit due for all this, I should mention that when I left the United States in 1984, I came to the Max-Planck Institut für Astrophysik, first on an Alexander Humboldt Senior Scientist award; I then stayed on as a guest scientist. I must acknowledge that if it had not been for the use of the facilities of the Institut, the hospitality, support and friendship of the researchers, I would not have been able to carry out the present work. It was my amazingly good fortune that many of the key, active objects I had observed with the big telescopes on the Pacific Coast were just being observed with the frontier-breaking X-ray telescope at the Max-Planck Institut für Extraterrestrische Physik (MPE). It picked out the most energetic objects with ease, and the telescope was still small enough so that it had a sufficiently large field to include the crucial objects which were related to the central progenitor galaxies.

All of the staff and faculty were enormously kind and helpful. To single out just a few: Rudi Kippenhahn, who initially nominated me for the Humboldt award and arranged for me to stay on afterwards; Hans-Christoph Thomas in the neighboring office, who was always ready to assist me in complex computer problems; and Wolfgang Pietsch at MPE, who taught me what rudiments of X-ray image processing I was able to learn, and showed me his many observational breakthroughs. We all have our precious beliefs, and the greatest courage is to respect a differing belief. Here I found people who believed the way one did science was the overriding ethic, and, with poetic justice, I think it leads to the greatest advances.

The following book is arranged with the first two chapters establishing that high redshift quasars emerge from the active nuclei of nearby galaxies. The next two chapters show that smaller companions of nearby galaxies also have intrinsic (non-velocity) redshifts, which persist down to the stars and gas that make up the galaxy. Chapter 5 discusses how the Local Supercluster is composed of similar groups and types of objects, and shows how their intrinsic redshifts decrease from the quasars down to the oldest galaxies. Chapter 6 introduces the startling evidence that faint groups of high-redshift, non-point-source objects on the sky are generally not distant clusters of normal galaxies, but instead are more like smaller, intrinsically redshifted components of broken up quasars.

Chapter 7 discusses how gravitational lenses cannot explain the association on the sky between quasars and lower redshift galaxies. It presents arguments that the quasars are not lensed background objects but younger material actually emerging from the central object. Chapter 8 presents the evidence for quantization, a phenomenon that could not occur if redshifts were caused by velocities. Chapter 9 discusses the theory. It points out how the Friedmann/Einstein expanding universe (the so-called "Big Bang") is based on a mistaken assumption—and why it cannot explain the observations. A more general solution of the basic equations is presented and it is discussed how it predicts the observed creation of quasars and their evolution into normal galaxies.

Finally, Chapter 10 recounts a number of examples where Academic Science has been unable to modify its theories and commitments to accommodate new observational facts. Directions of possible change are briefly discussed.

But the text, I feel, is not as important as the pictures. If non-specialists find parts of the text too technical, it is recommended just to scan through these sections. Actually, the pictures tell the story. One can look at some of the key pictures and simply understand by analogy with everyday experience the important aspects of how objects are related to each other, and how they must develop with time. In fact, the whole book could be reduced to a few pictures in which a person's ability to recognize patterns and sequences would convey most of the meaningful information. If individuals have confidence in what they "see," they can live serenely with the knowledge that they do not yet have ultimate understanding.

Introduction

WHY ARE REDSHIFTS THE KEY TO EXTRAGALACTIC ASTRONOMY?

Redshifts: If we look at the light from an object after it has been spread out from short to long wavelengths, we will see peaks and valleys due to emission and absorption from its atomic elements. One thing we can then measure is how much these features are displaced from their wavelengths in a laboratory standard.

It turns out when we observe galaxies and quasars, such features are generally shifted to longer wavelengths, in some cases by amounts up to 4 or 5 times the local laboratory values. This redward displacement of lines in the spectrum is considered to increase with distance and to be the most significant information we have about the faint smudges that are supposed to represent the most distant objects we can see in the universe. But if the cause of these redshifts is misunderstood, then distances can be wrong by factors of 10 to 100, and luminosities and masses will be wrong by factors up to 10,000. We would have a totally erroneous picture of extragalactic space, and be faced with one of the most embarrassing boondoggles in our intellectual history.

Because objects in motion in the laboratory, or orbiting double stars, or rotating galaxies all show Doppler redshifts to longer wavelengths when they are receding, it has been assumed throughout astronomy that redshifts always and only mean recession velocity. No direct verification of this assumption is possible, and through the years many contradictions have arisen and been ignored. The evidence presented here is, I hope, convincing because it offers many different proofs of intrinsic (non-velocity) redshifts in every category of celestial object—from stars through quasars, galaxies and clusters of galaxies. Moreover, this one key observable will ultimately lead us to consider a universe governed by the non-local effects of inertial mass and quantum mechanics, rather than the local dynamics of general relativity.

Cosmology: Because it concerns our ultimate origins and our future destinies, most people are interested in the nature of the universe in which we live. We call this picture of our environment in its broadest possible sense cosmology.

There is now a fashionable set of beliefs regarding the workings of the universe, greatly publicized as the Big Bang, which I believe is wildly incorrect. But in order to enable people to make their own judgments about this question, we need to examine a large number of observations. Observations in science are the primary and final authority. In the present book I endeavour to discuss these observations in as much detail as necessary to understand them. If the basic data were not so fiercely resisted by conventional cosmologists, the details would not need to be extensively discussed. But as it is, each block in the edifice has to be defended against endless objections. Moreover, the link between many different results is what ultimately gives the whole new picture credibility. The separate observations have to be related to each other, and this takes some patience and effort, although it is exciting to see the pieces fit together in the end. In order to make this process more stimulating, I recount some of the personal and human reactions that accompany these events. This, I hope, will aid the reader in understanding not only the facts, but why they have been received as they have. After all, science is a human undertaking, and people will only read the detailed scientific evidence if someone speaks freely about what it means in the context of real human beings.

Academia: Experts in physical science now are almost exclusively trained in universities. Our society financially supports theoretical scientists and facilities primarily through the academic hierarchy.

So there is another reason why it is not sufficient to relate just the new factual results. The current beliefs are the crowning achievement of our research and learning institutions, and if they are so completely wrong—and have been for so long in the face of glaring evidence to the contrary—then we must consider whether there has been an overwhelming breakdown in our academic system. If so, we must find out what went wrong and whether it is possible to fix it.

In order to put the pertinent observations into their proper perspective, I present the following table, which gives a loose outline of modern cosmology:

TABLE I-1: KEY EVENTS IN COSMOLOGY

1911	W.W. Campbell	redshifts of OB stars (K effect)
1922	A. Friedmann	solution of Einstein's field equations
1924-	E. Hubble	island universes and
1930		redshift relation
1948	J. Bolton	double lobed radio sources
1963	Palomar	Quasars
1970's	G. de Vaucouleurs	Local Supercluster
1980-	Satellite	X-rays
1990's	observatories	Gamma rays
	Cosmic Ray Telescopes	Ultra High-energy Cosmic Rays

Future:	Redshift as a function of age
	Quantization of redshift
	Episodic creation of matter
	Mach generalizes Einstein
	Mass as a frequency resonance

Key Events in Cosmology—The Theory

It is currently believed that rigorous cosmology started in the early 1920's after Einstein wrote down the equations of general relativity. These essentially represented the conservation of mass, energy, momentum, *etc.* in the most general possible coordinate system. In 1922, the Russian mathematician, A. Friedmann, "solved" these equations, *i.e.*, showed how the system would behave in time. It is interesting to note that at first, Einstein felt this solution was incorrect. Later he said it was correct, but of no consequence. Finally he accepted the validity of this solution, but was so unhappy with the fact that it was not a stable solution, *i.e.*, it either collapsed or expanded, that he retained the cosmological constant he had earlier introduced in order to keep the universe static. (This constant was later referred to as the cosmological fudge factor.)

In 1924, Hubble persuaded the world that the "white nebulae" were really extragalactic, and a few years later announced that the redshifts of their spectral lines increased as they became fainter. This redshift-apparent magnitude relation for galaxies became known as the Hubble law (through lack of rigor, often referred to as the redshift-distance relation). At this point Einstein dropped his cosmological constant as a great mistake, and adopted the view that his equations had been telling him all along, that the universe was expanding. Thus was born the Big Bang theory, according to which the entire universe was created instantaneously out of nothing 15 billion years ago.

This really is the entirety of the theory on which our whole concept of cosmology has rested for the last 75 years. It is interesting to note, however, that Hubble, the observer, even up to his final lecture before the Royal Society, always held open the possibility that the redshift did not mean velocity of recession but might be caused by something else.

Key Events in Cosmology—The Observations

In 1948, John Bolton discovered radio sources in the sky. Martin Ryle, a reigning pundit, argued furiously that they were inside our own galaxy. Of course they turned out to be overwhelmingly extragalactic. The curious thing was that they tended to occur in pairs, and it was soon noticed that there were galaxies between the pairs. I remember the noted experts of the day assuring us that the pairs had nothing to do with the galaxies.

Then radio filaments were found to connect these pairs (they later came to be called radio lobes) to the central galaxies, which were generally weaker radio sources. Without ever raising a glass of champagne, people began to think that they had always known that the radio sources were ejected in opposite directions by some explosive activity in the central galaxy. This fundamentally changed our view of galaxies: rather than vast, placid aggregates of majestically orbiting stars, dust and gas, it became clear that their centers were the sites of enormous, variable outpourings of energy. Probably this change of concept has still not completely sunk in for many astronomers. It is

Fig. I-1. The radio galaxy Cygnus A showing ejection of high
energy, radio emitting material in opposite directions from the
central object. This map was measured at 5 GHz with the Very
Large Array at Socorro, New Mexico by Rick Perley.

astonishing to note how closely the X-ray pairs across galaxies now correspond to the
ejected radio pairs, and how stubbornly people refuse to accept them as ejected.

Figure I-1 here shows a radio image of one of the first double-lobed radio galaxies
discovered, Cygnus A. Seeing the thin jets leading out into the swept-back lobes leaves
no room for doubt that this is a result of ejection from the central object. Something
initially small and associated with radio emission has had to come out from the center of
this galaxy. Quasars are also often radio sources, and many examples will be shown in
this book of pairs of high redshift, radio- and X-ray emitting objects, obviously ejected
from active central galaxies. The reason, of course, for the rejection of the pairing
evidence for quasars is the now-sacred assumption that all extragalactic redshifts are
caused by velocity and indicate distance. The association must be denied because the
quasars are at much higher redshift than the galaxies from which they originate.

Quasars

In 1963, some radio sources which had been identified with apparent stars were
being studied spectroscopically. What were puzzling stellar spectra, however, suddenly
turned out to be emission line galaxy spectra shifted to very long wavelengths. There
was some hesitation at first about accepting these redshifts as due to recession velocities
that approached the speed of light, since this would indicate great distance. At their
redshift distances, these objects had to be 1000 (and in the end 10,000) times brighter
than previously known extragalactic objects. But no other redshifting mechanism was
deemed likely, and everyone soon got used to these extraordinary luminosities.

Although the radio positions came from various observatories, the spectroscopic identification was done mostly at the Palomar 200-inch reflector. I was observing at Palomar at the time, but the positions were distributed privately. So I instead undertook a multi-year study of peculiar galaxies with the aim of studying how galaxies were formed and evolved. When the *Atlas* was complete, I discovered that across my most disturbed peculiars were pairs of radio sources. Very nice. Obviously the disturbance had been caused by the ejection of the radio sources. Then came the shock: some of radio sources turned out to be quasars! And the galaxies were not at great distances, but relatively close by.

Suddenly it is 30 years later; I am living in Germany and observing by satellite (computer processing the data) and writing about all the exciting new pairs of X-ray quasars across active galaxies which are being discovered by the ROSAT satellite telescope whose headquarters are located in the institute next door. There is only one little flaw in my idyllic good fortune, and that is that there is a relentless effort to ban all these lovely new observations from conferences and suppress them from publication. The compensation is that a few courageous, and officially disparaged, scientists are meeting and communicating with one another to explore the fundamental meaning this new information holds out to us.

High-energy Radiation and the Local Supercluster

Since the 1980's not only have satellite telescopes been telemetering down X-ray data, but more recently higher energy gamma-ray data has been gathered, and now ultra high cosmic ray detectors on earth have been reporting even higher energy radiation. In a separate chapter to follow, we will discuss the concentration of this energy at the center of our Local Supercluster and its possible meaning. But first, we should briefly describe the Local Supercluster because, contrary to common belief, this may be the only region of the universe we know much about.

The empirical results of galaxy catalogues were already showing in the 1950's that galaxies were not distributed uniformly over the sky. Yet the analyses by Gerard de Vaucouleurs showing the distribution along the supergalactic equator and the concentration around the Virgo cluster at the center were privately ridiculed, until suddenly in the early 1970's everyone discovered that they had known about it all along. It turns out that we will find the oldest galaxies there—and the most energetic radiation—perhaps pointing to current matter creation. Virgo may thus be a very special place in terms of understanding what we can currently see of our universe.

Future Events

At the bottom of Table I-1, some current investigations are listed. The investigation of redshift as a function of age already started in the early 1970's; quantization of redshifts shortly thereafter; and the creation of matter, perhaps in the 1980's. Since even the existence of these effects is not accepted at present; we can only say that they are epochal science in the making, if they are someday accepted.

Quantization of redshift and episodic creation of matter combine to offer the most promising empirical understanding of extragalactic objects, as explained in the following chapters. As a capsule preview of how galaxies are born, we can say that they are ejected from older galaxies as compact objects with low particle masses. As these newer galaxies age, and grow in size and mass, they in turn eject newer generations in a cascading process. We can actually show in Chapter 8 how groups of a dozen or so active quasars fraction into more and more objects, which in turn eventually evolve into clusters of large numbers of galaxies. The redshifts, which are very high as the newly created matter emerges from its zero-mass state, continue to diminish as the mass of the matter grows. Discrete steps in the redshift values are present throughout, but grow smaller when the overall redshift grows smaller. These aggregates of matter develop into normal galaxies, much like our own and those around us in the Local Group and Local Supercluster. All of this is almost diametrically opposed to the conventional view of galaxies condensing out of some tenuous, homogeneously pervading hot gas. It is a process that is going on in our own Local Supercluster, and, contrary to what is claimed by the Big Bang theorists, we do not know much about what may exist at cosmic distances. It turns out that for what we currently see, but do not understand, the essence is in the changes it is undergoing.

The final possibilities for a more fundamental understanding of the nature of matter as a function of frequency and time will be discussed at the end of the book. A complete understanding might be the ultimate reward for a careful analysis of all the observations. It is clear, however, that if we are to make progress in this area, we cannot wait for establishment science to, perhaps, someday accept the empirical results.

The Stars in 1911

When the first telescopes were being built under clear skies and systematic spectroscopic observations started—for example with the 36-inch refractor at Lick Observatory on Mount Hamilton—it was natural to observe what one could. That meant bright stars. One of the things that could be measured accurately was line shifts in stellar spectra. As the data accumulated, it was noticed that the bright blue (OB) stars, the hot luminous stars, had lines which were slightly, but significantly, shifted to the red. In 1911, the Director of the Observatory, W.W. Campbell gave the enigmatic name "K effect" to the phenomenon. (Actually K represented the expansion term in the formula that described the motion of all the stars measured.)

Since all the other stars in our galaxy moved together in reasonable ways, it was *not* concluded that we lived at the center of an expanding shell of OB stars. The effect was unexplained until the 1930's, when Robert Trumpler again found the effect in clusters of young stars in our galaxy. He thought he could explain it with a gravitational redshift at the surface of these hottest, most luminous stars. But that failed when the surface gravity turned out to be too weak. Later Max Born and Erwin Finlay-Freundlich tried to explain it with tired light. But that did not catch on. So the observations were again buried and forgotten.

I think it is a supreme and delicious piece of irony that 85 years after the Director at Lick Observatory announced the K effect Margaret Burbidge, a senior professor at the University of California, went up Mount Hamilton on a winter night to that same Lick Observatory. She observed two quasars that all the biggest and most advanced telescopes in the world had deliberately refused to look at, and in so doing, solved the riddle of the K effect—and at the same time laid the last flower on the grave of Big Bang cosmology.

Looking back now, especially from the standpoint of the coming chapters, we can see that if the relativists had heeded the published observations, going back a decade before their theoretical revelations, perhaps they would have decided that the universe was not necessarily exploding away from us in all directions.

My career at the Observatories in Pasadena slightly overlapped Edwin Hubble's. He personally gave me my first job: to aid in determining the crucial distance scale in cosmology. As a result I lived for two years on Mt. Wilson measuring novae in the Andromeda Nebula (M31). I moved on to observe Cepheid variables from South Africa and, finally, am now presenting evidence for a much different, perhaps truer, distance scale at greater distances derived from quasars and young galaxies.

In his seminal book *Realm of the Nebulae* Hubble wrote: "On the other hand, if the interpretation as velocity shifts is abandoned, we find in the redshifts a hitherto unrecognized principle whose implications are unknown". In the ensuing years the evidence discussed in the present book has built up to the point where it is clear that the velocity interpretation can now be abandoned in favour of a new principle which stands on a firm observational and theoretical foundation.

After about 45 years, I now know that if the academic theoreticians at that time had not forced his observations into fashionable molds, we might at least not have started off modern cosmology with the wrong fundamental assumption. We could be much further along in understanding our relation to a much larger, older universe—a universe which is continually unfolding from many points within itself.

Chapter 1

X-RAY OBSERVATIONS CONFIRM INTRINSIC REDSHIFTS

Just another isolated case. Your eye slid over that phrase because you wanted to see whether the referee was going to recommend publication. The answer was: not for the *Astrophysical Journal Letters*. The message behind the smooth, assured phrases was clear: "No matter how conclusive the evidence, we have the power to minimize and suppress it."

What was the evidence this time? Just two X-ray sources unmistakably paired across a galaxy well known for its eruptive activity. The paper reported that these compact sources of high-energy emission were both quasars, stellar-appearing objects of much higher redshift than the central galaxy, NGC4258. Obviously, they had originated from the galaxy, in contradiction to all official rules. Slyly, the referee remarked that "because there was no known cause for such intrinsic, excess redshifts the author should include a brief outline of a theory to explain them."

My mind flashed back through 30 years of evidence, ignored by people who were sure of their theoretical assumptions. Anger was my only honest option—but stronger than that provoked by many worse "peer reviews" because this was not even my paper. I did not have to stop and worry that my response was ruled by wounded personal ego.

How did this latest skirmish begin? Several years earlier an X-ray astronomer had come into my office with a map of the field around NGC4258. There were two conspicuous X-ray sources paired across the nucleus of the galaxy. He asked if I knew where he could get a good photograph of the field, so he could check whether there were any optical objects which could be identified with the X-ray sources. I was very pleased to be able to swivel my chair around to the bookshelves in back of me and pull

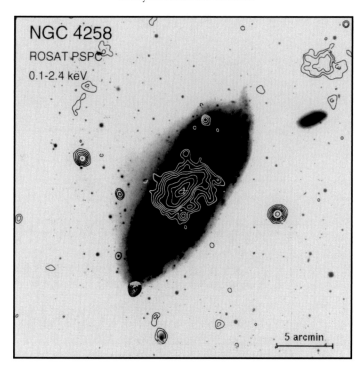

Fig. 1-1. The Seyfert galaxy NGC4258 is known to be ejecting material from an active
nucleus. A deep photograph is shown here with contours of X-ray emission
superposed (W. Pietsch et al.) The two point sources of X-rays on either side of the
nucleus coincide with blue stellar objects (BSO's).

out one of the best prints in existence of that particular field. I had taken it with the Kitt
Peak National Observatory, 4-meter telescope, about a dozen years previously. It was
very deep, because I had been searching this active galaxy for low surface-brightness
ejection features and associated high redshift objects.

Wolfgang Pietsch quickly found a small pointing correction to the satellite
positions and established that his X-ray pair coincided with blue stellar objects at about
20th apparent magnitude. (Figure 1-1) At that instant I knew that the objects were
almost certainly quasars, and once again experienced that euphoria that comes at the
moment when you see a long way into a different future. In view of the obvious nature
of these objects I felt Pietsch showed courage and scientific integrity in publishing the
comment: "If the connection of these sources with the galaxy is real, they may be
bipolar ejecta from the nucleus."

Then the dance of evasion began. It was necessary to obtain optical spectra of the
blue stellar candidates to confirm that they were quasars and ascertain their redshifts. A
small amount of time was requested on the appropriate European telescope. It was
turned down. Pietsch's eyes avoided mine when he said "I guess I did not explain it
clearly enough." The Director of the world's largest telescope in the U.S. requested a
brief observation to get the redshifts. It was not done. The Director of the X-ray

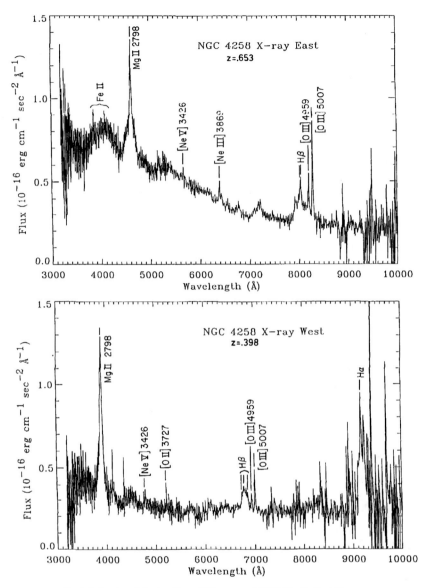

Fig. 1-2. Spectra of the two X-ray BSO's across NGC4258 with the Lick Observatory 3 meter telescope taken by Margaret Burbidge showing the similarity of the quasars.

Institute requested confirmation. It was not done. Finally, after nearly two years, E. Margaret Burbidge with the relatively small 3 meter reflector on Mount Hamilton, on a winter night, against the night sky glow from San Jose, recorded the spectra of both quasars. It was fortunate that mandatory retirement had been abolished in the U.S., because by this time Margaret had over 50 years of observing experience.

Of course, the referee report from which I quoted was directed against her paper, which reported this important new observation. In her firm, but lady-like English way, Margaret withdrew her paper from the *Astrophysical Journal Letters* and submitted it to the European journal *Astronomy and Astrophysics Letters*.

What was particularly appalling about this series of events was that Margaret Burbidge was someone who had given long and distinguished service to the scientific community. Professor at the University of California, Director of the Royal Greenwich Observatory and President of the American Association for the Advancement of Science among other contributions. It seems it was permissible to let her fly anywhere in the world doing onerous administrative tasks, but her scientific accomplishments were not to be accorded elementary scientific respect and fair treatment.

Some would argue that this is a special case, owing to the climate of opinion where the offices of the *Astrophysical Journal Letters* are located. But, as events in the following chapters make clear, the problem is pervasive throughout astronomy and, contrary to its projected image, endemic throughout most of current science. Scientists, particularly at the most prestigious institutions, regularly suppress and ridicule findings which contradict their current theories and assumptions.

Since scientific research in the end is almost completely supported by public funds, it behooves us as citizens to be aware of whether this money is spent wisely in relation to the real needs of, and possibilities for, the future of the society. The central purpose of this book is to explore this topic, and we will return often to it. But in the case at hand, the greatest progress can be made by discussing the actual observations of how nature works and the ways which science often misinterprets and misrepresents them.

Just another Experimentum Crucis—NGC4258

The referee's unconscious satire, "just another isolated case" was accompanied by deprecatory remarks such as "the quasars are not that well aligned" and "they are not exactly spaced across the nucleus of the galaxy." Of course a normal person would simply glance at the pair of X-ray sources across NGC4258 and realize they were physically associated. The average astronomer, however, would look at them and start to argue that they must be accidental, because astronomers now feel compelled to fit the observations to the theory and not *vice versa*.

Consequently, to head off the derogatory rumors which pass for scientific evaluation, someone had to compute a numerical probability. Basically, this meant computing how dense the associated X-ray sources are on average over the sky at a given apparent brightness. Then I had to ask myself: What is the chance of a source of a given brightness falling this close to an arbitrary point in the sky? Given the chance that the first one falls accidentally as close as the measured distance, then one must multiply by the chance that the second one falls at its observed distance. (*I.e.* if one out of ten will have a source as close as the real case then only one out of 100 will have two such sources.) For the two sources across NGC4258 it turns out that this chance is 5×10^{-2}

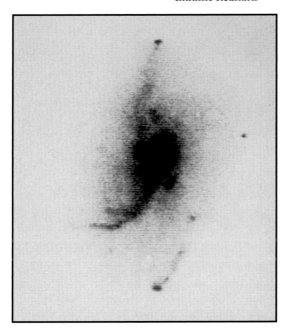

Fig. 1-3. NGC4258 photographed in the light of hydrogen alpha emission showing excited gas emerging from the nuclear regions (P. Roy *et al.*).

(*i.e.* five chances in one hundred). Of course, this does not include the improbability that they would be aligned across the nucleus of NGC4258 to within 3.3 degrees out of a possible 180. Nor does it include the improbability that they would be so equally spaced across the nucleus. Nor does it include the similar strengths and energy distributions of the two sources (which would not be expected from random, unrelated sources). Altogether the chance of this pairing of X-ray sources across NGC4258 being accidental is only 5×10^{-6} (5 chances in a million).

When it was confirmed that both sources were quasars, their redshifts became available. It was immediately apparent to anyone experienced with quasar spectra that these two were unusually similar. (Figure 1-2) A conservative probability for this similarity can be estimated at 0.08. Therefore the total probability of this association, being accidental, became $< 4 \times 10^{-7}$ (less than 4 in ten million).

Scientists claim that for acceptable scientific rigor, numerical probabilities must be calculated. But no matter how intimidatingly complex the calculation, no matter how small the probability of accident may be, the calculation does not tell you whether the result is true or not. In fact, no matter how significant the number is, scientists won't believe if they don't want to. When I submitted the paper with the calculations on NGC4258 which were claimed to be scientifically necessary, it was not even rejected, just put into an indefinite holding pattern and never acted upon to this day.

In the case of NGC4258, however, most astronomers overlooked a very important fact—that this galaxy is not just another object in a sea of identical objects. It is one of the most active nearby spiral galaxies known. In fact, in 1961 when the French astronomer G. Courtès discovered glowing gaseous arms emerging from the center of its concentrated Seyfert nucleus (see Figure 1-3), it led to observations with the

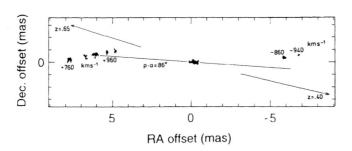

Fig. 1-4. Spots of water maser emission in the innermost nucleus of NGC4258 (M. Miyoshi *et al.*) showing approximate alignment in direction of quasars and redshift differences of the order of ± 1000km/sec correlated with the ejection velocities of the quasars.

Westerbork radio telescope which revealed that these proto spiral arms were also sources of synchrotron radiation (high-energy electrons spiraling around magnetic lines of force). In the past I had argued with Jan Oort, the discoverer of rotation in our own galaxy, about whether spiral arms were caused by opposite ejections from active nuclei. NGC4258 was the only case in which he ever admitted that proto-spiral arms were being ejected from the center.

The simplest and most obvious conclusion was that the pair of X-ray quasars was also being ejected from this unusually active galaxy. Interestingly enough, shortly after the discovery of the quasar pair, it was discovered that water masered (emission from H_2O molecules) spots in the inner .008 arc seconds showed redshift deviations of plus and minus ~1000 km/sec from the redshift of the nucleus. A conventional model explained this to be caused by a rotating black hole of 40 times greater mass than even the largest previously hypothesized. But just a glance at the observations (Figure 1-4) showed what instead looks like entrained material pointing out in either direction roughly toward the quasars. Note the quantitative agreement with the conclusions of van der Kruit, Oort and Mathewson (*Astronomy and Astrophysics.* 21, 169, 1972): "...clouds expelled from the nucleus in two opposite directions in the equatorial plane about 18 million years ago, at velocities ranging from about 800-1600 km/sec."

Even if the conventional hypothesis of black holes were tenable, and the hypothesized mechanism of bipolar ejection valid for them, the observation would still testify to the extreme activity of NGC4258, and thus support the association of the quasars. I personally prefer the concept of a "white hole", a place things irreversibly fall out of rather than into. For me, the whole lesson of the *Atlas of Peculiar Galaxies* was that galaxies are generally ejecting material. The merger mania seems to be a first guess based on a cursory look at galaxies. But I also think that the observations are not yet detailed enough to suggest a specific mechanism of ejection. Instead, the startling evidence of the association of high redshift quasars with low redshift galaxies needs to be faced. Those observations are more likely to lead to an understanding of the ejection mechanisms when responsibly pursued.

Of course, the evidence of association has been implacably rejected for 30 years by influential astronomers. In the case of NGC4258 just described, the chance of accidental association is only one in 2.5 million! A reasonable response would be to notice such a case and say, "If I see a few more cases like this I will have to believe it is real." Most astronomers say, "This violates proven physics [*i.e.* their assumptions] and therefore must be invalid. After all, no matter how improbable, it is only one case." Then, when they see another case they treat it *de novo* and reject it with the same argument. Professional scientists, however, have a responsibility to know about previous cases. And they do. When they block them out, it is a clear case of falsifying data for personal advantage—a violation of the primary ethic of science.

In a more general perspective, it can be said that the unique capability of human intelligence is pattern recognition. It is the most difficult task for a computer to perform. When one thinks about it, indeed, the seminal advances in science, and perhaps human affairs in general, were made by recognizing patterns in natural phenomena.

Other Cases of Galaxy-Quasar Associations

In order to forestall this argument about NGC4258 being "just another isolated case," I realized that it was more or less up to me to try to publish a paper which established its relation to other similar cases. It would also be necessary to calculate numerical probabilities in each case. As mentioned before, this is an obligatory exercise that critics do not like to do themselves, but insist on in any discovery paper. After the ritual argument about statistics is finished—is it 10^{-5} or 10^{-6}?—the argument is sufficiently abstract that people who wish to disbelieve the result can ignore it, when in fact it would be embarrassing to ignore it as a straight judgment call from looking at a picture.

Of course, the most important purpose was to gather together more examples of the same kind of pairing. That should clinch the issue. Table 1-1 is reproduced here from the paper that never appeared in *Astronomy and Astrophysics*. Already at that time, it showed five cases involving only X-ray pairs of quasars around lower redshift objects, each with a chance of less than one in a million of being accidental.

Table 1-1. Some X-ray Pairs Across Galaxies

Central Gal	z_G	r_1	r_2	$\Delta\theta$	$F_{x,1}\times10^{-13}$	$F_{x,2}\times10^{-13}$	z_1	z_2	p_1	p_{tot}
NGC4258	.002	8'.6	9'.7	3°	1.4 cgs	0.8 cgs	.40	.65	5×10^{-2}	$< 4 \times 10^{-7}$
Mark205	.07	13'.8	15'.7	44°	2.3	2.7	.64	.46	2×10^{-2}	connected
PG1211+143	.085	2'.6	5'.5	8°	.2	1.4	1.28	1.02	1×10^{-2}	$< 10^{-6}$
NGC3842	.02	1'.0	1'.2	33°	1.0	.3	.95	.34	7×10^{-5}	6×10^{-8}
NGC4472	.003	4°.4	6°.0	1°	~3000	~800	.004	.16	2×10^{-4}	$< 10^{-6}$

Subscript 1 designates nearest source; $\Delta\theta$ represents accuracy of alignment; F_x's are estimated for .4 – 2.4 keV band except last entries which refer to M87 and 3C273 and are HEAO 1, 2–10 keV band; $p1$ designates accidental probability of finding the sources of strength F_x at r_1 and r_2; $1 - p_{tot}$ gives estimated probability of physical association.

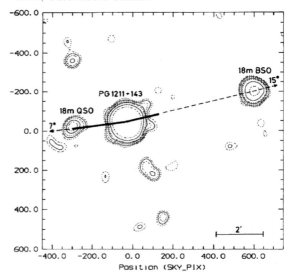

Fig. 1-5. X-ray map of the
Seyfert/quasar
PG1211+143. The X-ray
BSO to the west was
confirmed as a quasar by
cooperative effort between
Beijing Observatory and
Indian Astrophysics
Institute in Pune. Solid line
shows this pair of quasars
coincides with line of radio
sources which would be
conventionally accepted as
having been ejected.

PG1211+143

One of the cases in Table 1-1 became available in the following way: During the time it took for the redshift measurements on the NGC4258 quasars to unfold, I traveled to the National Radio Observatory and gave a talk. Afterwards, Ken Kellermann came up to me and said, "Here is a bright quasar which appears to have a line of radio sources passing through it—and one of the radio sources is a higher-redshift quasar." As soon as I returned to my office in Munich I asked my friend and computer expert in the next office, H.C. Thomas, to show me how to search the archives for any X-ray observations of this object. (After one year all proprietary observations are put in a public archive—but considering the amount of specialized knowledge one needs to access these records, the term "public" is rather euphemistic.)

An observation was found, and I eagerly reduced the approximately 4 megabytes of data to form an X-ray picture of the field. As Figure 1-5 shows, the central object is strong in X-rays, and the radio quasar to the East is conspicuous. But most electrifying, there is the hoped-for, strong X-ray source just on the other side, to the West. I immediately went to the Sky Survey photographs and found that this latter X-ray source coincided with a blue, stellar appearing object (BSO). Another pair of quasars across an active object! And this one was aligned with radio sources which by now are accepted as ejected from active galaxies!

But now the same old problem, how to obtain a confirming spectrum and get the redshift? Big observatories were obviously out of the question. The quasar candidate was rather bright, however, and it probably could be observed with a smaller telescope. I sent the finding charts to Jayant Narlikar, Director of the Inter University Center for Astronomy and Astrophysics in Pune, India. He interested a young researcher in obtaining a spectrum with the 1 meter Vainu Bappu telescope. The observation was scheduled in April, however, and the monsoon moved in. Despair—it was gone for the year!

Jayant said that they had asked the Beijing Observatory to do it, but I did not take that seriously, because to my knowledge, China did not have adequate equipment. It turns out, however, there is some reason to look forward to e-mail, because a month later I was delighted to receive a message that the spectrum had been obtained by the Chinese. After some normalizing of photon counts, it was possible to derive a redshift of $z = 1.015$.

The confirmation of this X-ray BSO as a quasar was particularly compelling, because PG1211+143 had been noticed as a result of its having a line of radio sources across it. As we have discussed, flanking radio sources are customarily interpreted as arising in ejection processes. How else could this pair of X-ray quasars, along exactly the same line, have arisen?

The numerical value of this redshift also turns out to be an important result. When included in Table 1-1, it showed that the difference of redshifts between the quasars in the first three, best pairs was .25, .18, and .26. In other words, interpreting the quasars as ejecta, the projected ejection velocities should be .082c, .058c and .060c, in km/sec. The coincidence of three independent determinations giving closely the same ejection velocity is very encouraging for this interpretation. (Velocities can only be added as in $(1 + z_i)(1 + z_v) = (1 + z_t)$ where i = intrinsic, v = velocity and t = total, as described in Chapter 8.) For an average projection angle of 45 deg., this gives an average true ejection velocity of .094c or 28,200 km/sec.

Radio Pairs from 1968

Several months after submitting this result, and in the midst of dealing with the usual hostile referee and nervous editor, I recalled a surprising fact. Back in 1968 I had investigated pairs of radio sources in the sky, some of which had turned out to be quasars.* From the estimated age of conspicuous disturbances in the central galaxy, and the measured separation of the quasars from their galaxy of origin, I had calculated ejection velocities of .1c. *In fact, I had calculated ejection velocities only five years after quasars had been discovered by a completely different method which now agreed well with the new measures!*

Of course, even so early in the game, such a storm had been raised against local quasars that there was no chance of publishing in a normal journal. As a result, I had published in the Journal of the Armenian Academy of Sciences, *Astrofyzika*. Viktor Ambarzumian was a hero of science in Armenia. We agreed on his initial insight that galaxies were formed by ejection from older galaxies. He did not believe my evidence at that time that redshifts were not velocity indicators. But as a tribute to his fairness, he did not hesitate for a moment to welcome my paper. The 12 figures in that paper are dramatic proof that the X-ray results of 1994 had been predicted in detail by the radio

---

* This reversed the original discovery procedure of 1966. Instead of finding pairs of high redshift radio sources across disturbed galaxies, I looked for pairs of radio sources on the sky and then looked to see whether there were galaxies between them. Of course, many of the radio sources in the pairs turned out to be quasars.

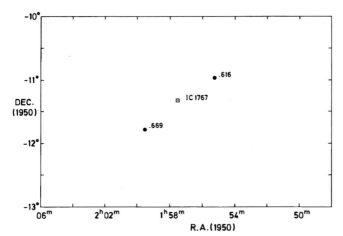

Fig. 1-6. The two strongest radio sources in the pictured area fall across the disturbed spiral galaxy IC1767. The redshifts of these radio quasars at z = .62 and .67 are so close as to confirm their physical relation. (H. Arp, *Astrofizika*, 1968).

quasars in 1968. The paper was also testimony to the fact that sensible analysis of observations was being blocked and ignored, while the high profile journals were submerged with a flood of elaborations of incorrect assumptions which prevented anyone from remembering anything important for more than a few years.

Figure 1-6 here shows a quasar pair from that 1968 paper. The pair involves the brightest radio sources in the field, and is so conspicuous that it is difficult to entertain any idea that it is an accident. Then, of course, there is the disturbed galaxy IC1767 falling at the center of the pair: how likely is that to be an accident? I did not know the redshifts when this pair was published in *Astrofyzika* but they were subsequently determined to be .67 and .62. Finally, out of a possible range for radio quasar redshifts from $.1 < z < 2.4$, what is the probability of getting two unrelated quasars to have redshifts within .05 of each other? This result was then published in *Astrophysical Journal*, but with the same lack of result. In the face of 28 years of accumulated evidence, to go on proclaiming that quasars are out at the edge of the universe seems unpardonable.

Markarian 205

The second entry in Table 1-1, which yields a projected ejection velocity of .058c, is from a famous and controversial association of a quasar-like object (Mark205) with a violently disrupted spiral galaxy (NGC4319). It is featured in color on the cover of my book *Quasars, Redshifts and Controversies,* and the long campaign to disprove the connection between the two objects is described therein. The connection was first shown in 1971, but as late as August 1995, there was still an exchange of letters in *Sky and Telescope,* in which one of the original disputants continued to claim the bridge did not exist.

The observations listed in Table 1-1 involve two new quasars and connections which were discovered in 1994. In 1990, the Max-Planck Institut für Extraterrestrische Physik (MPE) launched the X-ray telescope ROSAT (Röntgen Observatory Satellite Astronomical Telescope). Actually the telescope, a superior work of engineering, was

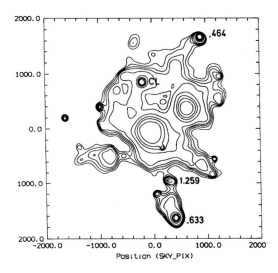

Fig. 1-7. X-ray filaments emerging from the Seyfert galaxy Markarian 205 and ending on quasars of redshift z = .46 and .64 (very similar to the z = .40 and .65 pair across NGC4258). This observation is shown on the front cover of this book and also in color Plate 1-7.

launched by Delta rocket from Cape Kennedy for which (plus one instrument) the U.S. claimed 50% of the observing time, leaving Germany with 38%, and Britain with 12% for building a small ultraviolet camera. Observing time was assigned to proposals from each country by allocation committees in that country. By an enormous stroke of good fortune, I was then a member of a German Institute and could submit proposals to the German selection committee. Even though I had been in Europe for four years, I still heard from friends in the U.S. how my previous requests for time on ground based telescopes and current space telescope requests had fared. Some secondhand accounts reported intense anger and ridicule expressed from the select group of the most reputable (but generally anonymous) astronomers who comprise the U.S. allocation panels.

My proposals to the German committee were rated very low, but at least the case was not hopeless. On the first few schedules I only received time for one very harmless proposal. But some of the experts at the MPE, located next door, were very helpful in preparing the proposals in the most acceptable possible format, and in later scheduling periods I received time on some "hot" objects. One of these was Mark205.

The proposal was to see if the connecting bridge from NGC4319 to Mark205 showed up in X-rays. As is often the case, the major aim failed. (I think now the connection to the galaxy is too old to show well in high-energy). *But what did show up was two X-ray filaments coming out of either side of Mark205 and ending on point-like X-ray sources (Front cover of this book, Figure 1-7 and color Plate 1-7)!* I immediately got out the sky survey prints and superposed scaled X-ray maps to see if they were optically identifiable. Lo and behold! They were not only blue stellar objects, but unusually bright in apparent magnitude.

Of course, they were quasars: but how to get the requisite spectroscopy which would give the redshifts? The same old problem, all the telescopes were occupied studying *distant* high redshift objects. Then a routine check of catalogued quasars bore unexpected good fortune. It turned out that a team of researchers had previously investigated fields around strong X-ray objects and found an excess of sources around Mark205. The excess sources turned out to be mostly higher-redshift quasars, but they rejected the significance of this on the basis that Mark205 had been previously known to be in an active region (!??!). I could almost forgive them for that inverted logic, because I was so happy to see the spectroscopy of the sources in the field. It turned out that there were three(!) confirmed quasars in the X-ray filament that I had discovered coming out of Mark205 in the ROSAT observations. The two major ones at the ends of the two filaments are listed in Table 1-1 with their redshifts, which yield a projected velocity of .06c each (average deprojected velocity .08c). As previously remarked, this now becomes a very important confirmation of the ejection velocities computed for radio quasars 27 years earlier.

But, of course, the stunning aspect of the ROSAT observations was that two quasars of redshift .63 and .45 are actually physically linked by a luminous connection to a low redshift object of z = .007. When I showed this to the local experts, there were alarmed stares followed by annoyance. "Of course, if you go faint enough you will find noise features or instrument imperfections which connect everything together." The frightening aspect of this reaction was that they were saying: "If the connection between these objects cannot be attributed to noise, there must be something wrong with the instrument." The latter possibility, even the mention of it, is enough to freeze any member of a well-funded project in his tracks.

Of course, I made the argument that since the filaments from Mark205 were sufficiently broad, coherent features, they obviously could not be noise. I also reduced an exposure of a bright X-ray star in the same way as Mark205, and showed that the faintest levels exhibited no imperfections, but just broke up into random noise as expected. Any non-expert would simply have reasoned that instrumental defects would not likely originate just from an active object, and certainly there would be no reason for them to end on the quasars in the field.

Nevertheless, it was clear that the best possible presentation of the data needed to be communicated. The communication was not easy. Both initial collaborators opted out, because I mentioned the word "ejection" in connection with the filaments ending on quasars. This was just before the word was mentioned in connection with the pair of X-ray sources across NGC4258, which later turned out to be quasars. Actually I became somewhat worried that the pair across Mark205 was not better aligned. In attempting to account for this I pointed out that the connecting filaments started out from Mark205 in initially opposite directions, but that the N one then curved over to the quasar in the NW. It was not until a few years later that I realized the Narlikar/Das model of ejected quasars, which required the increasing mass of the ejected object to slow its initial high velocity, fitted the X-ray observations around Seyfert galaxies very well. Then the light went on: the N quasar on its way out had been gravitationally attracted to the companion galaxy NW of NGC4319 which had swung it around in the observed direction.

But the referee complained because the data tables were not arranged in a certain order, and the objects were not discussed in a certain sequence, and it had not been "proved" that the connections and extensions were not noise. The inevitable ritual was upheld, and the paper was stalled indefinitely.

The IAU Symposium

Fortunately, the International Astronomical Union (IAU) was holding its three-yearly meeting in Holland in August 1994. A four-day symposium on Examining the Big Bang and Diffuse Background Radiation had been appended. Now my participation was always a matter of doubt, but this time not enough members of the organizing committee spoke against it to prevent my being invited to give a short paper. I realized I could cram most of the important new observational data on the new cases of X-ray quasars associated with low redshift galaxies into the five pages of a camera ready paper. Even though it would take more than a year to appear in the little-read *Proceedings*, it was at least a publication to which interested researchers could be referred to see the vital pictures of the actual X-ray data.

Returning early from the peace of the family vacation in the French Alps, I picked up my transparencies and diagrams and headed off to entertain the power elite with deliciously forbidden "crackpot ideas." (The establishment always confuses data with theories.) There were a few other dissidents in attendance to whom it was very important to communicate the new observations. Jayant Narlikar gave a rigorous presentation of how, near mass concentrations, new matter could be "created" in the vicinity of old matter. Geoff Burbidge gave his usual pungent update of the evidence that some quasars were much closer than their redshift distance.

The symposium relentlessly advanced toward one of its high points. The customary authority on extragalactic theory was scheduled to give the inevitable summary of the present state of knowledge. It always pained me that even though everyone knew what was going to be said, it was given the better part of an hour, whereas the new observations which destroyed the premises and conclusions of the talk never had enough time to be presented in 15 or 20 minutes (and usually not at all). Clearly, the main purpose of these "review of the theory talks" was to fix firmly in everyone's mind what the party line was so that all observations could be interpreted properly.

The reviewer of choice was naturally Martin Rees—recently having glided effortlessly from Plumian Professor to Astronomer Royal of England. After the standard defense of the Big Bang (even though it did not need defending) the only substantive comment from the audience was from the perceptive veteran Prof. Jean-Claude Pecker. He pointed out inconsistencies in the use of galaxy evolution as an adjustable parameter in order to avoid unexpected behavior with redshift in the Big Bang.

The final day consisted of a panel of about 9 members picked to represent the range of topics covered during the symposium. Facing the audience on the extreme right was Martin Rees, middle-left Geoffrey Burbidge and on the extreme left, myself. Rees opened up with a strong attack on the observations I had shown in my short talk a few days previously. When it came my turn to make an opening statement, I showed even more startling observational images that contradicted conventional models. The discussion was then thrown open to the rather large audience and a Dutch journalist, Govert Schilling, rose to ask Martin Rees a question.

The question, roughly paraphrased, was: "In view of the evidence Dr. Arp has shown, why have not major facilities been used to further observe these objects?" Martin turned toward me and erupted in a vitriolic personal attack. He said I did not understand the evidence from superluminal motions, that I did not believe the age of galaxies, plus a number of other elementary failings. I was rather stunned by the vehemence of this response, and I suppose the audience was also. After a moment or so, I replied that superluminal velocities were not a problem if you put things at their correct distances, and that I, of all persons, should believe the ages of galaxies, because as a graduate student I had measured the countless stars in globular clusters which helped establish the only age we have for galaxies. But most important of all I said, "I feel it is the primary responsibility of a scientist to face, and resolve, discrepant observations."

An Amateur Observes Mark205

What apparently set off Rees in response to the journalist's question was that it had been mentioned that an amateur had observed the NGC4319-Mark205 connection with the Hubble Space Telescope. Since 1971 this had been considered a crucial object in the proof of discordant redshifts of quasars, and in the symposium I had shown new evidence for the association of further, higher-redshift quasars with the same system. Because the Space Telescope was reputed to be able to answer all questions, many people had urged us to observe this key object again. Jack Sulentic, a long time collaborator on this project, and I prepared a complex, time consuming observing proposal—the kind that automatically sifts out the outsiders. It was not only turned down, but savaged by the allocation committee. So much for that exercise in futility. I was informed later in a letter that "it was NASA's policy not to release the names of scientific assessment panels." My first image was of my colleagues in false beards and dark glasses sneaking into the meeting room. Then a less humorous thought occurred to me—that large amounts of public money were being handed out by a secret committee.

It was not long before a delightful story started to circulate. The Space Telescope administrators had decided to make 10% of the time available to the community of amateur astronomers. This is actually a well-deserved acknowledgment of an enthusiastic, knowledgeable and important community. The rumor was that they had asked to observe Mark205. I did not really believe this until several years later when the author of the proposal himself walked into my office and put the observations down on my desk. He was a well-informed and able high school teacher who had quite competently confirmed the bridge between the low-redshift NGC4319 and the high-redshift Mark205. I urged him to publish, but to this day I have not seen it in print and I do not know what difficulties he may have encountered.

As a side note: Someone observed the galaxy NGC1073 with the three quasars in its arms with the William Herschel Telescope in La Palma. I thought I saw some filaments associated with the quasars, but I have seen nothing published yet. Finally, one amateur was assigned time to observe spectroscopically the quasar which is attached by a luminous filament to a galaxy called 1327-206. But NASA set the Space Telescope on the wrong object! Shortly thereafter, the Space Telescope Science Institute announced it was suspending the amateur program because it was "too great a strain on its expert personnel."

An even greater embarrassment was, however, that all these objects were drawn from my book *Quasars, Redshifts and Controversies,* the contents of which the NASA allocation committees had been avoiding at all costs. As we will have occasion to mention a number of times during this book, amateurs have a much better grasp of the realities of astronomy because they really *look* at pictures of galaxies and stars. Professionals start out with a theory and only see those details which can be interpreted in terms of that theory.

This is some of the background behind the sensitive point which the journalist raised with Martin Rees in the final discussion panel. The reason the point is so sensitive is that the influential people in the field know what the observations portend, but they are too deeply committed to go back. The result will surely be to inexorably push academic science toward a position akin to that of the medieval church. But if that is the evolutionarily necessary solution, then perhaps we should hasten the process of replacing the present system with a more effective mode of doing science.

Table 1-2. Galaxy-quasar associations investigated in X-rays through 1995

Galaxy	Quasar	Redshift	Separation	Probability
Mark474/NGC5682	BSO1	z = 1.94	1'.6[1]	5×10^{-3}[2]
NGC4651	3C275.1	.557	3.5	3×10^{-3}
NGC3067	3C232	.534	1.9	3×10^{-4}
NGC5832	3C309.1	.904	6.2	7×10^{-4}
NGC4319	Mark205[3]	.070	0.7	2×10^{-5}[4]

1 Separation of quasar measured from NGC5682 nucleus.
2 Probability from Burbidge *et al.* (1971).
3 On cosmological hypothesis Mark205 is 0.5 mag. Less luminous than the definition of a quasar.
4 Probability that a Seyfert galaxy would fall within 0'.7 of an arbitrary point in the sky.

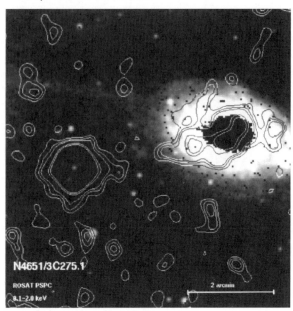

Fig. 1-8. The optical jet, spiral galaxy NGC4651 showing an X-ray jet emerging from its nucleus directly to the quasar with redshift z = .557. See also Fig. 7-11 for larger area view around the galaxy.

X-Ray Observations of Galaxy-Quasar Pairs

In addition, I reported for the first time in IAU Symposium 168, the results of pointed ROSAT observations on four additional galaxy-quasar pairs that fell conspicuously close together on the sky. (See Table 1-2). The probability of these associations being accidental was already very small back in the 1970's, and when the

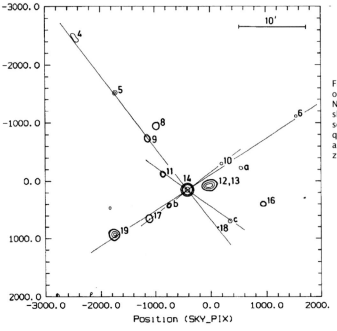

Fig. 1-9. An X-ray map of the area around NGC4651/3C275.1 showing lines of X-ray sources from the quasar. Source no. 4 is a catalogued quasar of z = 1.477.

observations revealed X-ray extensions from the galaxies toward the quasars, it not only clinched the physical association of these objects of vastly different redshift, but it also confirmed the ejection origin of the quasar from the galaxy.

One of these pairs, NGC4319-Mark205, has already been discussed here, but the others are mentioned below because of the understanding they add to the nature of the galaxy-quasar relation. The two most compelling cases are discussed first.

NGC4651/3C275.1

The radio-bright quasar 3C275.1 is situated only 3.5 arcmin from the bright apparent magnitude spiral galaxy NGC4651. The probability that this would occur by chance is only about 3 in 1000. But what no one ever calculated was the compound probability that the galaxy it fell so close to would be the one spiral galaxy in the brightest 7000 that had the most conspicuous jet emerging from it. That reduced the accidental probability to less than 1 in a million. Now Allan Sandage, who had photographed this galaxy in 1956, nervously grasped the implication, but immediately pressed the argument on me that the galaxy jet was not pointing at the quasar, which proved that it had nothing to do with the quasar. Of course, it was only pointing 20 degrees away from the quasar, and subsequent deeper plates revealed that there was material filling in under the jet, down to within a direction only 6 degrees away in position angle from the quasar. (See Figure 7-11 in a later chapter). But by that time the configuration had been relegated to the category of disproved associations.

Actually, there is an amusing story about the statistical association of the whole group of radio bright 3C quasars with bright apparent magnitude galaxies that G.R. Burbidge, E.M. Burbidge, P.M. Solomon and P.A. Strittmatter (B^2S^2) established. They found a less than 5 in 1000 chance of accidental association for the whole sample. When I showed the X-ray extension from the nucleus of NGC4651 almost to the position of the quasar to Prof. J. Trümper, the Director of the X-ray section of the Max-Planck Institut für Extraterrestrische Physik (MPE), I mentioned that this was one of a class of galaxies known statistically to be associated with quasars. He was very skeptical until I remarked that the B^2S^2 result had been confirmed by Rudi Kippenhahn (former Director of the Max-Planck Institut für Astrophysik). After that he wanted analyses only in the latter form! But as I brought more and more results to him he said, "Well I know you can't be right, but I will help you where I can." I had to ruefully admit that was not completely discouraging—in fact, it was about as as much encouragement as I ever got.

Figure 1-8 shows that X-ray material stretches from the nucleus of the galaxy toward the position of the quasar, where the quasar material extends almost to meet it. If the 10.5 kilosec exposure had been just a little bit longer, it might have shown the bridge to be continuous. But does that really matter, considering the low probability of accidental contiguity, the low probability of such an active jet galaxy being accidentally involved, and the vanishingly small probability that an X-ray jet would accidentally be coming out of the nucleus of the galaxy and pointing directly at the quasar? It would

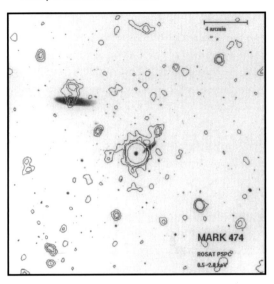

Fig. 1-10. The NGC5689 group, a typical association of active objects around a large, low redshift galaxy. X-ray contours show Markarian 474 to be a very active Seyfert with an X-ray filament leading out to a quasar with a redshift of z = 1.94. The companion galaxy, NGC5682, is just to the upper right of Mark474.

seem to me that a healthy science would eagerly recall all the other cases which pointed to the same conclusion and get on with the job of finding out why.

Figure 1-9 shows another characteristic tendency of these active objects, *viz.*, to exhibit lines of sources emanating from them. Also shown in this picture is a tendency of the lines to be nearly at right angles to each other—something we will see many times. The former is easy to picture in a model where the sources are ejected from active galaxies and quasars. A cause for the latter is difficult to imagine, but when we get a mechanism that gives such ejections, it may be a sign we are approaching understanding. (Robert Fosbury, the ESO expert on Seyfert galaxies, tells me the optical ejection cones from these active galaxies have de-projected opening angles of about 80°.)

Mark474 and the NGC5689 Group

This is a prototype of the groups which I believe represent the building block units which make up our known universe. Like the groups of galaxies we know the most about, such as our Local Group and the next nearest large group, the M81 group; the NGC5689 group has as a spiral galaxy like that of type Sb as its dominant galaxy. Actually NGC5689 is classified as an Sa; but it is the same morphological type of massive rotating galaxy with a large central bulge of old stars. (Figure 1-10)

My attention was first called to it by Edward Khachikyan, an Armenian astronomer friend. B.E. Markarian, another Armenian astronomer, had found this very high surface brightness, ultraviolet rich galaxy now called Mark474. Next to it was the lower surface brightness galaxy NGC5682, which turned out to be a companion to the large NGC5689 and, characteristically, had a redshift about 100 km/sec greater. Mark474 had a redshift about 10,000 km/sec greater. I felt that the companion should have an associated quasar, and looked on the Palomar Schmidt prints for a blue object in the

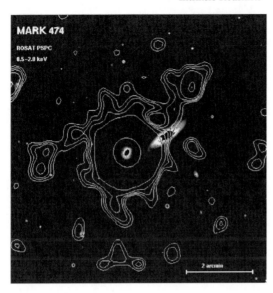

Fig. 1-11. A closer view of the Seyfert galaxy Mark474 showing the X-ray material connecting to the quasar at the upper right (small dot inside smallest contour). Note material extending from the quasar in a direction away from the Seyfert.

neighborhood. I found it, but it was a little too faint for the poor spectrograph on the 200-inch telescope. I asked Joe Wampler at Lick Observatory to get the spectrum, and it turned out to be a quasar of redshift z = 1.94. (Maarten Schmidt criticized me for going outside the Hale Observatories to get this spectrum on a smaller telescope, but I replied that Joe was the only one who had built a good enough spectrograph—the Wamplertron—to observe the object.)

Now I had a close triplet of unusual objects which were almost certainly associated, in spite of their vastly different redshifts. While I was sitting in an MPE working group, my ears perked up when I heard that this Markarian object had been discovered on the survey to be a copious source of X-rays. Arguing that such a strong source deserved to be observed in a pointed observation, I was able to obtain a 12,862 sec exposure in the low resolution mode. The initial reduction showed everything I had hoped for. The quasar was well visible in X-rays, and was connected back to, and elongated away from, the strong X-ray Seyfert. (Figure 1-10). (Actually it is unusual to see such a faint apparent magnitude, high redshift quasar detected in X-rays.)

Figure 1-10 also shows that X-ray emitting material is being ejected along the minor axis of the "parent" galaxy in the system, NGC5689. The interesting implication here is that even though the presently active galaxies in the group probably evolve rapidly into more quiescent entities, the original galaxy in the group is capable of subsequent ejection episodes. It is also apparent that there are other relatively strong X-ray sources aligned in an "X" pattern across Mark474. Most of them are identified with blue stellar objects (BSO's), and clearly represent additional quasars associated with this active group.

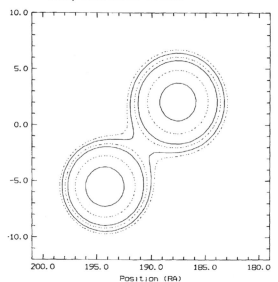

Fig. 1-12. Instrumental spread of photons around two unrelated point sources. Only outer isophotes hourglass together with inner contour lines returning quickly to symmetry.

The optimally smoothed X-ray contours are shown in enlargement in Figure 1-11. Skeptics immediately argue that if one puts two unrelated distributions of photons close to each other, they will meld together to form an apparent connection. Yet if one thinks about it for a moment, one realizes that they intermingle only in the outer isophotes to form an hour glass-like shape. Figure 1-12 here shows an isophotal contouring of two adjacent instrumental point-spread functions. It is clear that only the outermost isophotes merge into an hourglass shape, and all the inner isophotes immediately return to circularity. Real elongated inner isophotes, filamentary connections and jets look conspicuously different, regardless of added noise.*

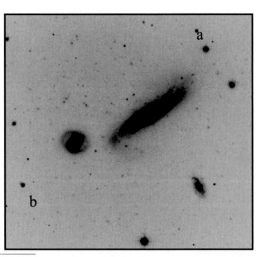

Fig. 1-13. X-ray source (b) discovered by H.G. Bi by deconvolving strong image of Mark474. The important result is the almost exact alignment of this blue, peculiar, X-ray object across the nucleus of the Seyfert with the quasar (a). What would a spectrum reveal about (b)?

---

*   Because of unfamiliarity with low surface brightness detection techniques coupled with non expectation of extended features, almost no use of such information has been made by X-ray observers. The X-ray archives are presently a gold mine of untapped information waiting for someone with access and computing power to harvest the data.

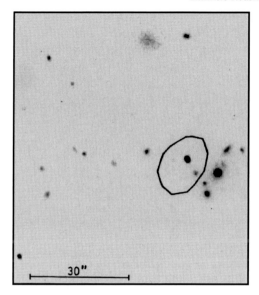

Fig. 1-14. Non-equilibrium configuration near Mark 474 of galaxies plus one X-ray BSO. What is redshift of the quasar candidate? What is nature of the extremely low surface brightness dwarf just to the north?

30"

In Figure 1-11 one can see that the connection between Mark474 and the quasar passes close to the companion galaxy NGC5682, the galaxy which I had originally felt was the origin of the quasar. Now, however, a collaborator in the office next to mine, H.G. Bi, applied a deconvolution program to the data in order to sharpen the resolution, and discovered a rather strong X-ray source buried in the outskirts of the Markarian galaxy. That X-ray source, was readily identified with a compact but deformed blue object. The decisive aspect then emerged—this new quasar-like object was almost exactly aligned with the known quasar across Mark474! (Figure 1-13). In view of the close pairings of quasars across Seyfert galaxies which have now emerged, this appears to be just another confirmatory pair across a Seyfert galaxy!

The NGC5689 group is typical also in the pattern of redshifts of the objects in the group. The largest galaxy has the lowest redshift; the smaller companion has a higher percentage of younger stars and a redshift hundreds of km/sec higher. There is a very active galaxy with thousands of km/sec higher redshift, and finally very high redshift quasars emerging from the latter. This theme is repeated over and over again. In later chapters, when we consider the redshifts decreasing as the objects age, we will try to suggest some possible reasons for this hierarchical, fractal pattern.

It would be helpful, however, to know the redshift of the compact blue object which is on the other side of Mark474 from the quasar, as shown in Figure 1-13. There is also an intriguing region situated midway between the dominant galaxy and Mark474. It consists of a string of red galaxies (a string being a non-equilibrium configuration which cannot last the age of the galaxies) containing an X-ray BSO. A peculiar dwarf galaxy is less than 1 arcmin away. The picture of this latter group is shown in Figure 1-14 and also in the publication of the proceedings of IAU Symposium 168 (ed. M. Kafatos and Y. Kondo). How long will it be before some of the numerous large

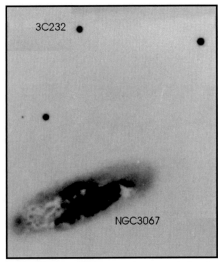

Fig. 1-15. Kitt Peak 4
meter photograph of
NGC3067 showing high
surface brightness and
shattered appearance of
absorption filaments.

telescopes around the world are used to observe these curious and intrinsically informative objects?

NGC3067 and the Quasar 3C232

This galaxy-quasar pair has had an absolutely amazing history. Back in 1971, Burbidge *et al.* derived a probability of accidental association of less than one in three thousand. A. Boksenberg and W.L.W. Sargent found absorption lines of the galaxy in the spectrum of the quasar in 1978 and assumed it was a distant, background quasar shining through the galaxy, a chance coincidence. In 1982, Vera Rubin *et al.* went further and attributed the spectral shift of the galaxy absorption lines to rotation around a massive galaxy taking place at an unusually large distance from its nucleus. Naturally, the latter calculation produced a mass of "dark" (undetectable) matter some 16 times the estimated mass of visible matter. Despite the enormity of this factor, it was hailed as proof of the existence of enormous amounts of unseen matter in the universe. But the galaxy was patently not an ordinary galaxy. It was a sharply bounded, very high surface brightness "star burst" galaxy—a rare and active kind of galaxy, which would make the

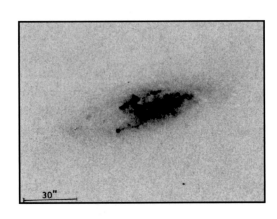

Fig. 1-16. Palomar 200-inch photograph of NGC3067 in light of hydrogen alpha emission showing ejected, hot gaseous filaments

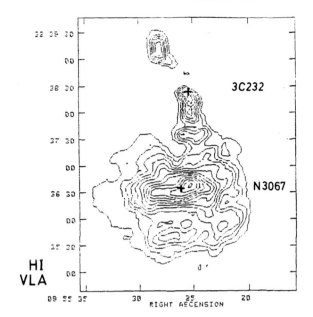

Fig. 1-17. Radio map of neutral hydrogen in NGC3067 showing filament leading from disturbed galaxy to quasar. Map from Carilli, van Gorkom and Stocke.

accidental association with a quasar hundreds of times less likely. Moreover, pictures of the galaxy revealed a shattered, explosive morphology with emission line filaments issuing from it (Figures 1-15 and 16). Under no circumstances could it be a normal galaxy in equilibrium rotation, which would be required in order to derive a meaningful mass. The huge derived mass was a complete fiction! Why didn't they look at the galaxy? (Actually I sent pictures, but to no avail).

An even more startling development occurred in 1989 when C.L. Carilli *et al.* found a filament of neutral hydrogen leading from the west end of the galaxy directly to and beyond the quasar! (Figure 1-17). The hydrogen had clearly come from the active galaxy—how else other than being pulled out by the ejection of the quasar? And notice that the quasar falls just at the densest point of the hydrogen distribution with contours of less dense gas trailing back towards the galaxy.

This extraordinary result should have cemented the association beyond any doubt, but later it was claimed that the configuration was merely accidental. J. Stocke *et al.* argued that the neutral hydrogen at the redshift of the galaxy absorbed continuum light from the quasar, but did not show excited optical emission lines, proving the quasar was quite far in back of the hydrogen filament. Because the other arguments are so overwhelming that we are dealing with another physically associated galaxy and quasar, I reread very carefully the complex calculations they had made. There it was: a "short" extrapolation. The photons they needed to ionize the hydrogen in the filament and make it fluoresce were at shorter wavelength than those in the spectrum. So they extrapolated to an unobserved portion of the spectrum. I extrapolated and got half their value. But regardless of how much quasar radiation was *extrapolated* to be shorter than this wavelength, the actual amount would be determined by the amount of hydrogen at

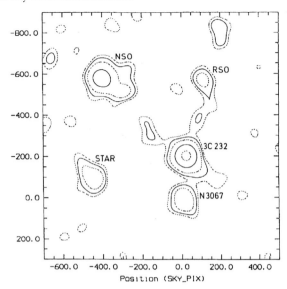

Fig. 1-18. Integrated, low surface brightness X-ray emission around the galaxy/quasar pair NGC3067/3C232. This represents another "cross" extension of X-ray material from active objects.

redshifts intermediate between the quasar and the hydrogen filament, the degree to which the filament was composed of small, dense clouds, and the relative beaming angle between the ultraviolet and radio wavelengths from the quasar. If the conventional paradigm had instead required the quasar and filament to be adjacent, which of these plausible configurations would have been announced as a new "discovery"?

The X-ray fun had only just begun. When the Einstein Laboratory Satellite went up, it observed the quasar because it was quite a bright object. At a workshop at the European Southern Observatory (ESO), I pointed out that there was an X-ray tail coming off the quasar in a direction opposite to the galaxy. Martin Elvis from the Cambridge Center for Astrophysics (CFA) jumped up and said, "That's noise." I argued that you could *see* that it was not noise. He said, "I'll look into it and report what I find." He never reported back.

When I got the relatively short 5600 sec exposure on it with ROSAT, there was the X-ray extension north of the quasar! In fact, there was another cross extension of X-rays (Figure 1-18)—quite similar to the configuration around Mark474. But the most exciting result was that there was a double-sided X-ray jet coming out of the nucleus of the starburst galaxy, NGC3067 (Figure 1-19). How many galaxies does one find with such conspicuous bipolar X-ray jets? When I showed this to my MPE colleagues they became angry at me for saying that I thought the jet was curving slightly around as it extended NE, even more toward the quasar, and that a longer exposure might show it leading directly to the quasar. Others said they thought the X-ray extensions from the quasar were just noise. Further X-ray observations on the object were rejected by the allocation committee.

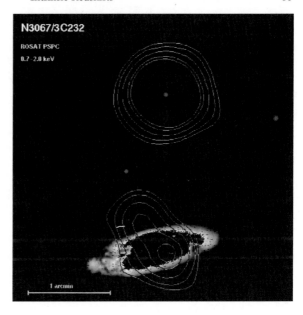

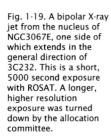

Fig. 1-19. A bipolar X-ray jet from the nucleus of NGC3067E, one side of which extends in the general direction of 3C232. This is a short, 5000 second exposure with ROSAT. A longer, higher resolution exposure was turned down by the allocation committee.

NGC5832 and 3C309.1

The last of the five pointed observations that I got with ROSAT was a very short exposure of 4300 sec on one of the Burbidge *et al.* galaxy/radio quasar pairs. Only the quasar registered, and the galaxy, relatively far away at 6.2 arcmin distance, did not. The distribution of X-ray sources in the field, however, was very interesting. As (Figure 1-20) shows there is again a strong line of sources running NE to SW through the quasar

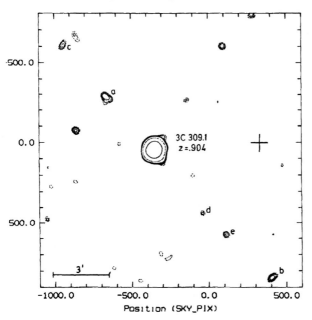

Fig. 1-20. A very short, 4300 second X-ray exposure shows only the quasar 3C309.1 but not the nearby galaxy NGC5832 (plus sign). Small X-ray sources in the field, however, form a line and possibly a cross through the quasar.

and the suggestion of a line coming off in almost an perpendicular direction. This configuration was criticized because some of the sources had only 3 to 6 counts. I argued in return that if the background is low enough, just a few counts make for significant sources, as can be well judged visually.

Chapter 2

SEYFERT GALAXIES AS QUASAR FACTORIES

Evidence that quasars were physically associated with low redshift galaxies had been amassing since 1966 (See *Quasars, Redshifts and Controversies* for the history through to about 1987). The following years saw further proofs accumulate, mostly from X-ray observations, and they are reported now in the previous chapter. But the stronger the evidence, it seemed, the more attitudes hardened against accepting these observations.

With the discovery of the pair of quasars across NGC4258, however, a new level of proof suggested itself. If more such striking pairs across active galaxies could be found, it would be hard to resist the ultimate conclusion. What more obvious class to inspect than those like NGC4258, namely Seyfert galaxies?

Seyfert Galaxies

The American astronomer Karl Seyfert discovered this class of galaxies in the 1950's by looking at photographs and noticing that some galaxies had brilliant, sharp nuclei. The emission line spectrum of such a galaxy signified that large amounts of energy were being released into its nucleus. For a long time, no one was worried where this energy came from. When the problem was finally realized, "accretion disks" came to the rescue—a kind of cosmic equivalent of throwing another log on the campfire. But the conspicuous emission lines did enable astronomers to do something they are good at—systematically classify and catalogue these objects.

Since Seyfert galaxies produced strong X-ray emission, by 1995 most of the brighter ones had been observed with the ROSAT satellite. This presented an opportunity to investigate a class of active galaxies which had been previously defined and more or less completely observed. The existing observations could be analyzed to

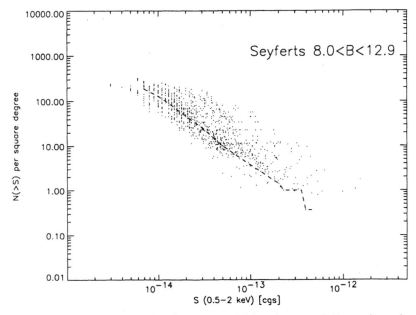

Fig. 2-1. Cumulative number of X-ray sources brighter than strength (S) around a nearly complete sample of bright Seyfert galaxies. Dashed line represents counts in non-Seyfert control fields.

see whether there existed more cases of pairs of quasars across active nuclei such as had been observed in the Seyfert galaxy NGC4258.

A search through the archived X-ray observations revealed that, of all Seyferts known, observations were 74% complete to 10th apparent magnitude and 50% complete to 12th. After some contaminated fields had been weeded out, a total of 26 fields were available.

Now came the formidable task of accessing and analyzing these fields. As mentioned previously, an enormous amount of specialized knowledge is required to enter the "public" archives. I found the perfect candidate to collaborate with me on this job, a German astronomer named Hans-Dieter Radecke. He had just finished doing a very important and courageous piece of work on the gamma ray observations in the region of the Virgo Supercluster which we will discuss later. But he was out of a job—and the problem was to find him some funding. It seemed hopeless, but as a last resort I asked Simon White, our new director at Max-Planck Institut für Astrophysik (MPA) if he could help. To our delight he found support for 6 months, and this made possible what I hope will be recognized as a crucial step forward in our understanding of physics and cosmology.

Hans-Dieter produced lists of sources, their strengths and positions, for each of the 26 Seyfert fields. Then, using exactly the same detection algorithm, he reduced 14 control fields. The control fields were within the same range of galactic latitudes and treated identically to the Seyfert fields. Therefore, when a significant excess of X-ray sources was found around the Seyfert galaxies, there was no question that these X-ray

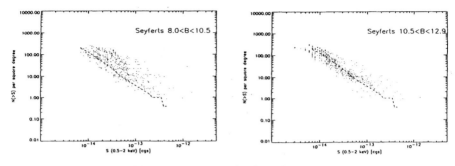

Fig. 2-2. X-ray sources around Seyferts of two brightness classes showing how associated sources become less bright as Seyferts become more distant.

sources belonged to the active galaxies. The sources were 10 to 100 times more luminous than sources usually found in galaxies, such as binary stars or supernova remnants, and they lay far outside the confines of the galaxy (generally from 10 to 40 arcmin away or several hundred kiloparsecs at the average distance of the Seyfert.) Practical experience guaranteed that these kinds of X-ray sources would be confirmed as quasars. *The beautiful feature of this result was that any astronomer could simply look at this one plot of X-ray strength versus number per square degree (as shown here in Figure 2-1), and realize that when these excess sources—which manifestly belonged to the Seyferts—were measured, they would almost all turn out to be quasars.*

With one economical diagram we had proved that Seyfert galaxies *as a class* were physically associated with quasars! This added enormously to the significance of the pairs across Seyferts such as NGC4258, because now the data were just telling us *in what way* the pairs were related to the active galaxy. In later sections we will discuss the obvious relation of the quasar pairs to pairs of radio sources which, since the 1950's, are acknowledged to have been shot out of the nuclei of active galaxies. Of course, in this sample of 24 Seyferts (omitting the brightest two as being too close to fit into the average sample—see Figure 2-2), many more pairs of X-ray sources were found. All told, there were 21 pairs of X-ray sources involving 53 BSO's (some pairs or alignments involved multiple X-ray sources). Almost every Seyfert had a pair of BSO's, most of which were certain to turn out to be quasars! Before we discuss some of these new pairs, however, it is interesting to comment on how these developments were received.

Spreading the Good News

Astronomers are always holding meetings, and as the journals become choked with papers, the meetings are increasingly the forum where new results are communicated (except for press releases, which are so hyped that they have to be heavily discounted). The meetings are traditionally the places where power relations are straightened out. It is painful for me to attend them because there is almost total conformity to obsolete assumptions. But I am old-fashioned enough to believe that when truly important new results come along, the conference organizers have a moral obligation to see that the results are presented.

With the new results in hand, I became optimistic that when they were communicated, they would finally persuade the researchers to at last begin to reappraise the fundamentals in the field. Many of the new results discussed in Chapter 1 were available at the time of the well known Texas Symposium on Relativistic Astrophysics, which was held in the adjacent institute in Munich in December of 1994. I submitted an abstract of a paper I wanted to give. The schedule was released, but my name was missing. About 14 years previously the Texas Symposium had been held in Munich, and I travelled all the way from California to give a summary paper on the evidence for associations of quasars with low redshift galaxies. The paper had made some impression then. But now, I was sad to say, after all this time, when the evidence had grown so much stronger, the newest evidence was not to be allowed. Sometime later, an international X-ray conference was held in the nearby town of Würzburg. Again I was excluded.

Finally, in 1996 I was awarded an Eminent Scientist invitation to come to Japan for three months. It was suggested that I could time my visit with an international conference on X-rays that was going to take place in Tokyo. The new results on the families of quasars around Seyferts were just out, so I sent in an abstract and arranged to come during that period. I was really joyous at the thought that this important information could be communicated in these circumstances, and that some sort of reconciliation could take place between people who were really interested in advancing knowledge. Just as I was packing, the conference schedule came out without my name on it.

Now, I am experienced enough to know how organizing committees pick speakers for conferences. And I have a rough idea of who, particularly in the most advanced countries, exerts pressure to keep what they consider rival research off the programs. But I am extremely saddened to realize that the members of the local organizing committees give in to such imperialistic pressure.

A Striking New Pair

It was exhilarating just to scan through all the new, good looking pairs of X-ray sources across the Seyferts in the maps obtained from the archives. A sample of these maps is shown here in Figures 2-3,4 and 5. Note that the X-rays are plotted just as received, and not averaged to give their mean position, which is on average accurate to a few seconds of arc. Even though the images enlarge as they occur further off axis, their disposition with respect to the central Seyfert and the relative X-ray brightness of the sources are conveyed very clearly.

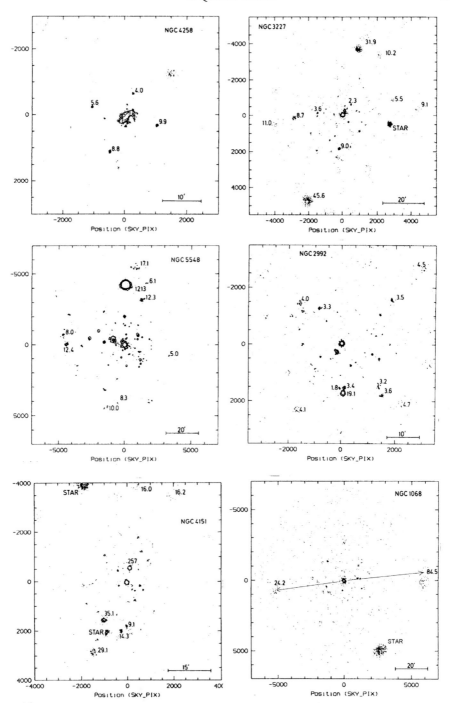

Fig. 2-3. A sample of the pairs of X-ray sources discovered across bright Seyfert galaxies. X-ray photons are plotted as received so that spreading of images with increasing distance from field center is conspicuous. Numbers represent counts per kilosecond. Lines in NGC1068 field represent direction of the distribution of water maser sources in the inner nucleus

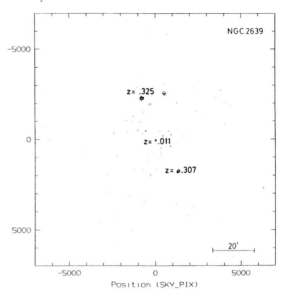

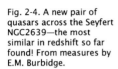

Fig. 2-4. A new pair of quasars across the Seyfert NGC2639—the most similar in redshift so far found! From measures by E.M. Burbidge.

An outstanding pair was immediately noticed across NGC2639. The two X-ray sources were both very strong (26 and 38 counts per kilosec). The identifications with BSO's were accurate and unambiguous (actually, one was a catalogued quasar that I had discovered near a companion to NGC2639 in 1980). Again, there was the need to get a spectrum of the other member of the pair—again Margaret Burbidge to the rescue. That pair of redshifts turned out to be a lift-you-out-of-your-chair thrill! As Figure 2-4 shows, the redshifts were z = .307 and .325, a difference of only .018. This was the closest that any of the pairs had been in redshift. What was exciting about this, of course, was that two unrelated X-ray quasars had only about one chance in 100 of accidentally falling this close in redshift. That probability, times the vanishing probability of finding such a strong pair of X-ray sources across an arbitrary point in the sky, made the whole computation a waste of time—here was clearly another physical pair across a Seyfert.

Fainter sources can be seen in Figure 2-4 aligned opposite to the z = .307 quasar and extending toward the z = .325 quasar. With fainter isophoting on an enlarged view of this region, four BSO's are optically identified, and clearly will represent a trail of quasars leading out to the z = .325 quasar when confirmed (see later Figure 3-26).

Another Water Maser

While the quasar redshifts were being measured, word arrived that stimulated emission from H_2O molecules had been observed in the nucleus of NGC2639—the same water masering that had been observed in the nucleus of NGC4258. This meant that the two best known pairs of quasars across a Seyfert fell across the Seyferts known to have the strongest "black hole" activity. The water maser lines in NGC2639 were particularly variable, showing velocity drifts of about 7 km/sec in a year.

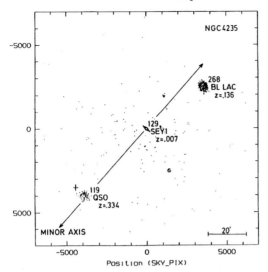

Fig. 2-5. Very strong (268 and 119 cts/ks) X-ray sources across the Seyfert NGC4235. Catalogued identifications as a quasar and a BL Lac object are labeled. Plus sign indicates the position of a Seyfert 1 of $z = .080$ identified previously but not registered in the present ROSAT X-ray map.

I had pointed out that the patches of water maser emission in the nucleus of NGC4258 were distributed approximately along lines in the direction of the two quasars (Figure 1-4 in preceding Chapter). So I was very pleased when Margaret and Geoff Burbidge wrote a short paper gathering together all the evidence for NGC4258 ejecting material in roughly these directions (in contradiction to the conventional interpretation, which had the rotation axis of the black hole at 90 degrees to this direction). Following this, *another* water maser was discovered in the center of NGC1068. As the lower right hand panel in Figure 2-3 shows, the orientation of the water masering spots (full line) *again* points in the direction of a strong pair of X-ray BSO's aligned across the nucleus of NGC1068. Now it may turn out that masering activity is common in Seyferts, as is ejection activity, but it also appears to be correlated with the strength or direction of the major ejection activity in the galaxy.

At the moment, the best guess as to what excites the water molecules is radiant energy in the beam associated with the ejection of the quasars. Why such a "cool" molecule is present in the very inner regions of such active galaxies may be a more challenging question.

A New Pair of Enormously Strong X-Ray Sources—NGC4235

When I first saw the X-ray map of the field around NGC4235, I was sure the pair of sources belonged to the Seyfert, because the chance of accidentally encountering such strong sources is only about 4 in 1000. Taking into account the alignment and equal spacing gives a total chance of the pair being accidental of only 6 in 100,000!

But I made a hasty assumption—namely that they were so strong in X-rays that they would be X-ray galaxies. So I only checked catalogues of known X-ray galaxies, and when I did not find them, I assumed they would have to be measured. After the paper was submitted I stumbled across the two sources catalogued; one as a very bright quasar of $z = .334$ and the other as a characteristically X-ray bright BL Lac object of $z = .136$.

(See Figure 2-5.) The discovery of BL Lac objects in associated pairs is extremely important. We will show in later sections that BL Lac's, because of their rarity, offer a powerful proof of associations, and therefore of intrinsic redshifts. They will also play an important role in the discussion of galaxy clusters in Chapter 6.

The Question of Ejection Velocity

In the first chapter we stressed the fact that the observations of pairs of quasars allowed us to compute a projected ejection velocity of about .07c. The NGC4235 pair just discussed would support this by giving a projected ejection velocity of .08c. (That is the intrinsic redshift of the quasar would be $z = .235$ but the velocity towards us would subtract $z = .099$ and the velocity away from us would add $z = .099$.) In Chapter 1, however, we showed one case where the redshifts in the pair were $z = .62$ and .67, yielding a projected velocity of only .015c. The separation on the sky for this case was about 1.3 deg., about 50% greater than other pairs associated with galaxies at this approximate distance from us. This made it plausible to argue that we were viewing the inevitable occurrence where the ejection was across our line of sight and the toward and away components of velocity were much reduced.

It was amusing to note that when the NGC4258 pair was first being discussed in a colloquium, Günther Hasinger challenged me to predict the redshifts of the probable quasars. The conventionalists clearly wanted a way to wriggle out of having to accept the association of the quasars. When they were measured at $z = .40$ and .65 I was encouraged that they were that close, and the conventionalists were relieved that they were not closer. But they should not have been relieved, because if quasars had been much closer, there would have not been enough velocity to get them out of the galaxy nucleus.

The pair across NGC2639 at $z = .307$ and .325, however, represents a more interesting situation. That would only allow .007c velocity component in the line of sight. For an ejection at 0.1c, we might expect to get such an orientation across the line of sight only about 9% of the time, a figure that can only be checked by measurements of many more pairs.

But there is another very interesting aspect to this problem. Do the ejection velocities represent a constant velocity of escape which will allow the quasars to pass out into the space between the galaxies; or do they decelerate as they reach larger distances from the ejecting galaxy? Do they keep going or stop?

The Narlikar-Das Model for Ejection of Quasars

By the 1980's I had produced strong statistical evidence that quasars were in excess density around younger companion and active galaxies. Jayant Narlikar and P.K. Das set out to make a dynamical model which could explain this. The problem was, assuming reasonable properties for the quasars, to find a way to keep the quasars in the spatial vicinity of the ejecting galaxy. Their model did this very nicely (*Astrophysical Journal*. 240,401).

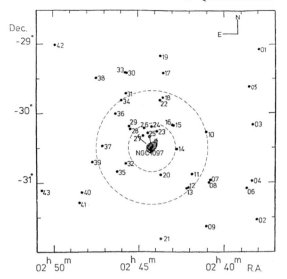

Fig. 2-6. All quasar candidates in a region around NGC1097 identified by X.T. He from a Schmidt, objective prism plate. The size of the PSPC and HRI fields investigated with X-rays are shown dashed.

One quantitative prediction of their model was that a quasar would reach a maximum apogee from the galaxy of about 400 kpc. Now it is very interesting that at the redshift distance of NGC2639, the two quasars are just about 400 kpc from the Seyfert. This would mean that the ejection velocity would have been lost and the quasars would be moving very slowly. Therefore quasars at larger distances from their galaxies of origin might be expected to have more closely matching redshifts regardless of the orientation of their ejection direction to the line of sight. Another aspect, which will be discussed in later chapters, is that quasars probably evolve to lower intrinsic redshifts as they age. In that case quasars of lower redshift would generally be expected to have smaller components of ejection or "peculiar" velocity—more like the galaxies into which they are evolving.

The Seyfert Galaxy NGC1097

NGC1097 has the most extensive, low surface brightness optical jets of any galaxy known. Plate 2-7 shows true color compositing by Jean Lorre of a set of the deepest 4 meter telescope plates ever taken at Cerro Tololo, Chile. On one side, just between the brightest optical jets, is a concentration of 5 or 6 bright quasars. These have been shown to represent an excess of a factor of 20 over expected background values, and about 40 quasars have been demonstrated to be concentrated around the galaxy (*Quasars, Redshifts and Controversies* pp 48-53 and 64). Figure 2-6 shows all the quasar candidates in the inner 2.85 × 2.85 degree center of an objective prism plate taken by the U.K. Schmidt telescope in Australia. The Chinese astronomer X.T. He picked these out by the appearance of their spectra; and in a two year observing program in Chile, I was able to verify with individual spectra that his accuracy of quasar identification was an amazing 94%. It is important to note that when this considerable work by a number of people

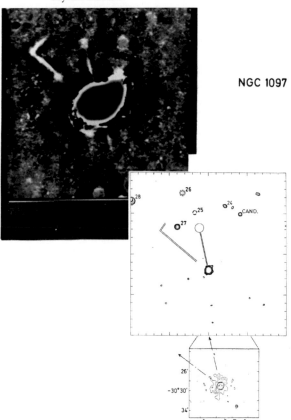

Fig. 2-7. Enhanced, star removed, composite (by Jean Lorre) photographs of NGC1097 showing luminous, crossed jets. Below center is the PSPC X-ray map of the field with known X-ray quasars numbered 24 through 28. Note faint, unidentified X-ray sources on the other side from the X-ray bright quasars. Bottom right shows an enlarged map in radio wavelengths with the two strongest jet directions marked.

NGC 1097

was published in 1983 and 1984, it already *established, at that time, the association of quasars around this one Seyfert galaxy that we are now finding to be characteristic of Seyferts as a class.*

In 1993 and 94, however, I received X-ray results of my own on this most exciting galaxy-quasar association. The data came from all three ROSAT modes, the low and high resolution pointed and survey modes. (The size of the fields covered by PSPC and HRI is shown in Figure 2-6.) When I first reduced the X-ray data, I was at once struck by the large number of X-ray sources in the field. Brighter sources in the NGC1097 field were more than 50% in excess of average control fields. The X-ray sources detected by ROSAT confirmed the earlier observations by the Einstein X-ray observatory and, in particular, confirmed that the brightest quasars fell just between and along the strongest optical jets. *Since it is difficult not to believe that the optical jets are ejected, it is obvious that the quasars are also ejected from NGC1097.*

These observations also showed lines and pairs of fainter X-ray sources coming out of the nuclear region of the Seyfert (Figures 2-7 and 2-8 and Plate 2-8). There was a large excess of X-ray sources around the disk region of the galaxy, and evidence for strong absorption of the soft X-ray component of many of the faint sources. Since it is known from optical studies of the galaxy that there is strong absorption in the disk of the galaxy (Figure 2-9), it was reasonable to suppose that the metals in this mixture of

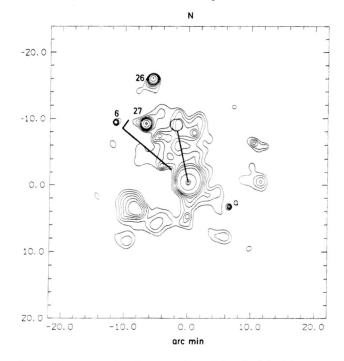

Fig. 2-8. The high resolution X-ray map (HRI) of NGC1097. Note material filling in toward the bright quasars 26 and 27. Note also the new point X-ray sources (6) and (a) aligned across the nucleus in the direction of the brightest jet.

dust and gas were also dimming the soft band of the X-ray sources. Unabsorbed, the X-ray sources would be bright enough so that they could reasonably only be quasars. The upshot is that these observations suggest many higher redshift quasars are being ejected, and that many may be encased in thick cocoons. Evidently, this is a busy quasar factory, and an interesting place to investigate in the far red and infrared.

We will see later that high redshift quasars ($z = 2$) are generally fainter than quasars in the $z = .3$ to 1.5 range. We have, however, already mentioned that the quasars

Fig. 2-9. Photograph of the barred spiral NGC1097 showing interior regions with opaque filaments and clouds of dust. A technique of emphasizing contrast in high surface brightness regions while preserving faint surface brightness features has been applied. (This technique was originally called automatic dodging but is now called unsharp masking.)

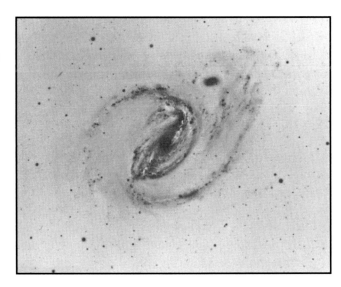

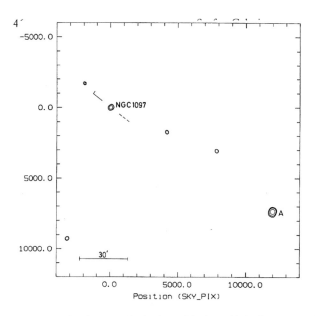

Fig. 2-10. An X-ray map from the all sky survey by ROSAT shows a very bright X-ray source about 1.9 degrees SW, roughly along line of the counterjet (dashed) to the brightest jet (full line). Spectrum by Tony Fairall identifies the source A as a BL Lac object.

appear to be born with high redshifts which decrease with age. Since many bright quasars with these smaller redshifts are associated with NGC1097; it is reasonable to suggest the fainter X-ray sources are high-redshift, younger quasars, many of which are just emerging from the central, dusty regions of the Seyfert.

For example, the high resolution X-ray observations (ROSAT HRI) show a pair of point sources paired across the NGC1097 nucleus (designated as 6 and a in Figure 2-8). These are not optically identified on deep, blue sensitive plates. They would probably be identifiable with the penetrating power of the infrared techniques on the new large aperture telescopes. What a useful project for these expensive facilities!

Figure 2-8 also shows low surface brightness material extending out from the nucleus of NGC1097, between the jets, to the location of two of the brightest quasars. It is not clear that this is X-ray material, because it does not show in the more sensitive PSPC observations. It is more probably ultraviolet light leaking in through an imperfectly blocked filter. (This possibility was doubted when I first published the evidence, but later a leak was verified in a measurement of the filter.) In any case the important aspect of this material is that it must arise from some form of hot gas which has been ejected along with the quasars! (One attempt to get spectra with the satellite ultra violet explorer failed on an administrative error, and the other attempt failed when a stabilizing gyro died.)

In the wide field of the ROSAT survey mode shown in Figure 2-10, there is a very strong X-ray source (A) identified about 1.9 deg to the SW. In fact, it is stronger than the very strong NGC1097 itself. It is identified with a bright (16.5 apparent mag.), probably stellar appearing object. Tony Fairall with the 74-inch South African telescope took a spectrum which demonstrated that it had a blue, continuous energy distribution, thus identifying it as a BL Lac object. This important kind of quasar-like object will be discussed immediately below. But first I would like to point to an important discovery in this survey X-ray field.

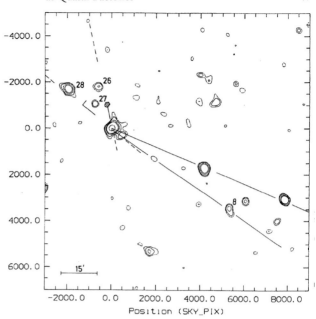

Fig. 2-11. An enlarged portion of the previous X-ray survey map shows sources along the general line to the BL Lac object. Perhaps of even more importance, the brightest quasar no. 28 lies along the line of the brightest jet and its X-ray contours are extended in both directions along the line of this jet!

As the enlarged Figure 2-11 shows, the BL Lac object is on a line of sources SW from NGC1097, which coincides very closely with the counterjet of the strongest optical jet to the NE. Along that major optical jet to the NE, however, is one of the brightest X-ray quasars belonging to NGC1097. *The X-ray isophotes of this quasar extend both backward and forward along the line of the strongest optical jet.* Since this alignment is obviously not an accident, and since the optical jet obviously originated by ejection from the active nucleus, this is *another* proof that the quasar must also have originated by ejection from the nucleus! Moreover, the strong X-ray BL LAC on the other side of the nucleus must then represent the other component of the ejected pair.

The Empire Strikes Back

Since the NGC1097 paper contained tables full of new source identifications from the analysis of the three different field sizes centered on the important Seyfert galaxy NGC1097, I thought it would be routine to publish in the journal which was carrying most of the European X-ray results of archival value. How wrong I was! The referee's report came back accusing me of "manipulating the data" and trying to claim an association of quasars with galaxies, which had "long ago been disproved." The editor forwarded these comments and rejected the paper on the grounds that he saw no need to reopen the debate. The extraordinary aspect was that four papers in addition to my own had just appeared in the same journal giving strong additional evidence for just such associations! The figures appear here for the first time, and the tabular X-ray data is still unpublished.

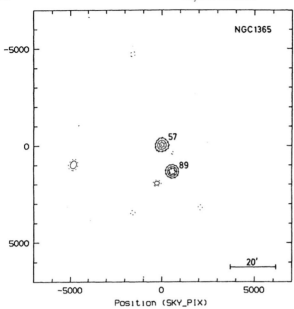

Fig. 2-12. X-ray map of the large southern spiral NGC1365. The strong source SSW of the Seyfert is a BL Lac object of redshift z = .308.

BL Lac Objects

These objects are so named because the prototype was originally classified as a variable star within our own galaxy. But then it was discovered that, in many cases, faint, redshifted lines could be detected on the strong continuum spectrum. Often these objects also showed faint nebulous edges to their images. The BL Lac's are now known to be strong radio and X-ray emitters, and are strongly variable.

They are also rather rare, and when they showed up in a ROSAT Seyfert field, they were very conspicuous because of their strong X-ray emission. Figure 2-12 shows an example of a BL Lac object close to the grand design spiral Seyfert NGC1365. While I was inspecting the 26 Seyfert archival fields discussed earlier, it was clear to me that the number of such objects encountered was significantly higher than would be expected in non-Seyfert fields. There is no real need to compute probabilities—but it can be done simply enough! The probability of encountering X-ray BL Lac's this bright, this close to the Seyfert ran from about 10^{-2} to 10^{-4}. (See Table 2-1). Therefore, the

Table 2-1 Current summary of BL Lac Objects in Seyfert fields

Seyfert	X-ray BL Lac	R	IPC F_x	P(BL)	V	z
Cen A	$(570)ctsks^{-1}$	114'	168×10^{-13}	1.5×10^{-3}	17.0 mag.	.108
NGC 1365	89	12	6.7	4.2×10^{-3}	18.0	.308
NGC 4151	257	4.5	14.8	4.1×10^{-4}	20.3	.615
NGC 5548	1213	35	(88.5)	4.1×10^{-3}	16	.237
NGC 4235	268	36	(19.6)	2.2×10^{-2}	16	.136
NGC 3516	156(Sl)	22	13.6	—	16.4	.089

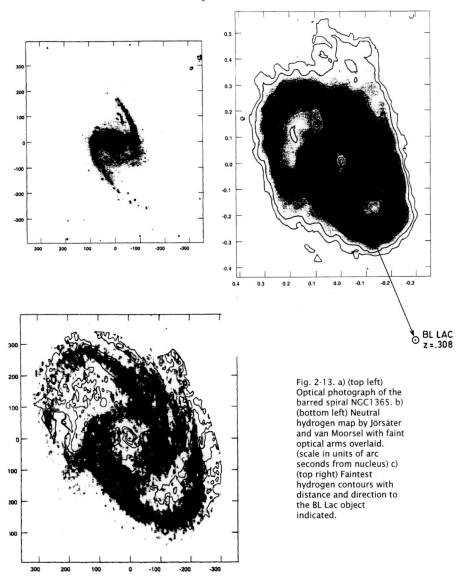

Fig. 2-13. a) (top left) Optical photograph of the barred spiral NGC1365. b) (bottom left) Neutral hydrogen map by Jörsäter and van Moorsel with faint optical arms overlaid. (scale in units of arc seconds from nucleus) c) (top right) Faintest hydrogen contours with distance and direction to the BL Lac object indicated.

chance of encountering the first five was about 3×10^{-5}, and if we count the object near NGC3516 as a BL Lac, a chance of only about 3 in ten million! A referee argued that due to the uncertain density of bright BL Lac's, this probability was uncertain—thus disproving the association! But even if it were only one in a million, the result is overpoweringly significant. Moreover the finding is restricted to just the five clear-cut BL Lac's encountered so far, and more are indicated in the full sample of 26.

"Ridiculous!" snorted the conventional astronomer. Who would believe a probability that small? Right! What is wrong? Well it's *a posteriori*, computed after you found the effect. So let's throw it out! Ah, but along came a great stroke of good fortune. In 1979, Jack Sulentic and I had tested the proximity of then known BL Lac's to bright apparent-magnitude, Shapley-Ames Catalogue galaxies and found an excess at a separation of about 1 deg. (the same as in the Seyfert fields). So the Seyfert result was not *a posteriori*, but a confirmation of a previously predicted result. The cautionary lesson here seems to be that no matter how significant the result, it is customary to try to invent a reason to discard it if it doesn't fit expectations. The game here is to lump all the previous observations into one "hypothesis" and then claim there is no second, confirming observation.

But most important of all: Does the result make sense? It does, and in fact it is expected on empirical grounds. Consider one of the quasars in the pair associated with NGC2639, which we just discussed. Its apparent magnitude was $V = 18.1$ mag. and its redshift $z = .307$. Compare that to the BL Lac object within 12 arcmin of the Seyfert NGC1365. That BL Lac object had an apparent magnitude of $V = 18.1$ mag. and $z = .308$. The X-ray flux from the quasar was strong, but the flux from the BL Lac object was 3.5 times greater—undoubtedly the signal of the strong non-thermal continuum which reduces the spectral lines characteristic of BL Lac objects to low contrast.

But BL Lac objects are notoriously variable. The implication, then, is that a BL Lac can turn into a quasar quite easily, and *vice versa*, since they are already very similar. The key point here is that BL Lac's are a rare kind of quasar which can be easily recognized because of their strong X-ray emission. Therefore, they are easily proved to be associated with active galaxies—confirming the proofs that the related kinds of objects, the quasars, are also associated.

It is interesting to inspect the neutral hydrogen maps of the grand-design, barred spiral NGC1365. Figure 2-13a shows the optical photograph. Figure 2-13b shows how the hydrogen concentrates in the spiral arms to the southwest of the galaxy. (One can see the multiple, ejected arms to the north of the galaxy which, at a glance, disposes of several decades of density wave theory for the formation of the arms.) But Figure 2-13c shows how this hydrogen is extended closely in the direction of the nearby BL Lac. In the following case of NGC4151 we will actually see a connection to a BL Lac.

The Seyfert Galaxy NGC4151

Another famous and extremely active Seyfert galaxy is NGC4151. A map of it and its surrounding companions is shown in Figure 3-18 in the next chapter. Presented here is the X-ray map in Figure 2-14, which shows that there is a line of X-ray sources stretching through the central active galaxy to the NNW and SSE. The two to the NNW are rather strong, at 16.0 and 16.2 counts per kilosecond, but they are relatively defocused, at 33.1 and 33.9 arc minutes from the center of the field. They therefore appear rather spread out. Now they, and the sources opposite them with 14.3 + 9.1 and 35.1 counts/ksec, are all identified with blue stellar objects (BSO's). Therefore, we have a case of two highly probable pairs of quasars aligned across this Seyfert, both are

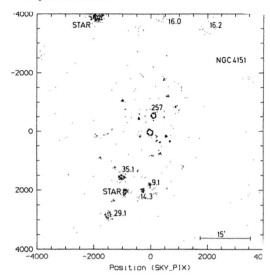

Fig. 2-14. An X-ray map of a 1.1 × 1.1 degree region centered on the large Seyfert NGC4151 (see Fig. 3-18). A strong BL Lac (257 counts/kilosec) is situated 4.5 arcmin N of the galaxy. Outer X-ray sources are also distributed generally along this line.

aligned fairly closely in the same direction. (One might also consider this one ejection with a narrow-opening cone angle.)

Along this line, about 4.5 arc min NNE from NGC4151, is a very powerful X-ray source, measured at 257 counts/ksec (compared to the Seyfert itself at 570 counts/ksec). This is a BL Lac object. Like the one falling next to the Seyfert NGC1365, it is very unlikely to have been encountered by chance. In this case, the probability is only 2×10^{-5} (see Table 2-1). But the object is also very unusual, in that it was first discovered in a radio mapping of the environs of nearby bright galaxies by the Westerbork telescope. Jan Oort had urged this project on me in collaboration with two Leiden radio astronomers, Tony Willis and Hans de Ruiter. We had identified this rather strong radio source in the NGC4151 field with a very faint stellar object. It was so faint that I had to use the multichannel spectrometer on the 200-inch at Palomar for several long exposures in order to try and determine its redshift. I only could register one line, and guessed that it was Lyman alpha at a redshift near $z = 2$. That turned out to be wrong, as John Stocke and collaborators later measured the object to have a redshift of $z = .615$.

The puzzle is this: What kind of object would be so faint optically and have such strong radio and enormous X-ray emission? As mentioned, it was highly probable that it belonged to NGC4151, and from Figure 2-14, it could be seen to lie in the apparent channel of ejection from that active Seyfert. But would there be any interaction in X-rays due to the spatial proximity of this strong BL Lac and the Seyfert? By searching the archives, Radecke and I found some HRI exposures of this field, and I set about looking at outer contours of the two images. *The outer, lower surface-brightness regions of NGC4151 revealed a filamentary extension which connected directly to the BL Lac object, as shown in Figure 2-15.*

Identifying luminous connections between objects of greatly disparate redshifts is a decisive way to establish their non-velocity character, as we have seen in the previous

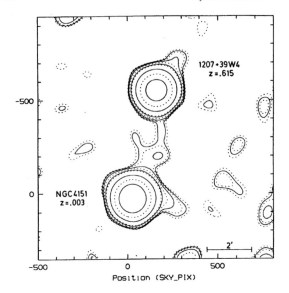

Fig. 2-15. High resolution X-ray map (HRI) showing low surface brightness connection between NGC4151 at z = .003 and the BL Lac at z = .615.

connections to quasars from Mark205, Mark474 and NGC4651. There could be a rich harvest of additional information if X-ray astronomers were to recognize the increased detection to be gained from integrating their data over extended surfaces. This is related to the old art of surface brightness photometry, but would require hiring people who were either experienced or motivated.

Finally, we call attention in Figure 2-16 to the innermost *radio* structure of NGC4151. At the high frequency of 5 GigaHerz, the resolution is so good that objects of less than ¼ arc sec can be seen emerging on a line on either side of the central nucleus, C4. X-rays cannot yield such high resolution, but show extension in the same direction. *Some compact, high-energy objects are being ejected in opposite directions from this compact nucleus—what else could they be but proto quasars?* This ejection direction obviously rotates with time, so only older ejection tracks would be pointing to outer, associated quasars. We will later grapple with the question of what state the matter is in when it first starts its journey, but the important inference for now is that small entities are ejected from the nuclei of active galaxies and evolve into high redshift quasars and allied objects.

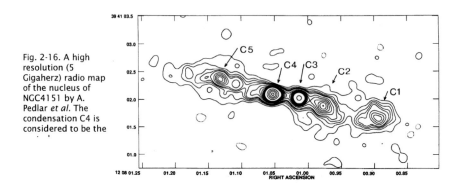

Fig. 2-16. A high resolution (5 Gigaherz) radio map of the nucleus of NGC4151 by A. Pedlar *et al.* The condensation C4 is considered to be the

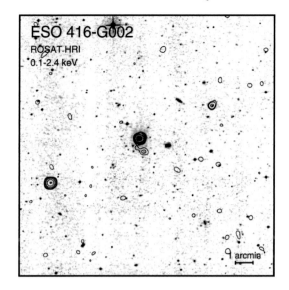

Fig. 2-17. The Seyfert galaxy ESO 416-G002 is about z = .03 redshift. The two aligned X-ray sources are identified with blue stellar objects whose spectra have only recently been observed. (Observations by Wolfgang Pietsch.)

ESO416-G002

During the time rumours were flying about Wolfgang Pietsch's pair of X-ray sources across NGC4258, the inevitable reaction was "Well that's just a curiosity; there won't be any more." But in addition to all the other cases described earlier in this chapter, in his own programs he had observed several other Seyfert galaxies which turned out to be just as devastating.

One is shown here in Figure 2-17. There are only three strong X-ray sources in the field. The source in the center is a Seyfert of z = 10,000 km/sec, and the two others, almost exactly paired across it, are accurately centered on stellar-appearing objects. Somehow, after more than two years of constant effort, it was never possible to obtain their spectra. Perhaps that speaks more eloquently than any further comment that could be made. (Recently Pietsch, with collaborators, confirmed the weaker of the pair as a quasar of about z = .6, and the stronger as a BL Lac object.)

Other Examples

Now that Seyfert galaxies have been identified as quasar factories, it is easy to look back and recognize all the other Seyferts which, in the past, were found to be the origin of associated quasars. Of the first two quasars to be associated with companion galaxies (see *Quasars, Redshifts and Controversies* pp 22-23), the quasar in the NGC5689 group turned out to be associated with the Seyfert Mark474 (see Chapter 1 here), and the quasar in the NGC7171 group turned out to be associated with a Seyfert. Mark205 is technically a Seyfert, although it is often called a quasar, and PG1211+143 is arbitrarily called a quasar, but very similar to Mark205 (redshifts z = .070 and .085 respectively).

There is a quasar GC0248+430 which—if you are ready for this—is described in the literature as "a possibly microlensed quasar behind a tidal arm of a merging galaxy."

The galaxy turns out to be a Seyfert 3, and the quasar has a redshift of z = 1.31. Many more of these galaxies associated with quasars may well turn out to be Seyferts when people get around to measuring them. That is not to say that only Seyferts eject quasars. Some good examples of "starburst" galaxies that give rise to quasars are NGC520, M82 and NGC3067. But then starburst galaxies are closely allied to Seyferts, and the classes may evolve rather rapidly. There is also the probability that outbursts occur intermittently, and after a galaxy has released some quasars it may become quiescent.

An example of an unworked gold mine is the starburst galaxy NGC7541. Described by Arp in the 1968 *Astrofyzika* article as being between a pair of bright radio sources, it subsequently turned out to have quasars of z = .22, .62, 1.05 and 1.97 around it. From the ROSAT survey, it has a pair of X-ray sources across it, and radio measurements to fainter levels show additional radio sources closely grouped around it. The main galaxy has a straight spiral arm, which looks like an ejection and has an early type stellar absorption spectrum. A close companion galaxy, NGC7537 appears active and might well be a Seyfert or allied type. This is the kind of region which requires a thorough observing program—the kind of program that used to be possible in the era of small telescopes, but is unthinkable in the era of big telescopes.

Summary of Empirical Evidence

In spite of a deliberate effort to avoid them, a large number of cases of quasars undeniably associated with much lower redshift galaxies have accumulated. Based on the discussion of the first two chapters of this book, the unavoidable conclusion, stated as simply as possible, is this:

It is clear that, spectroscopically, a quasar looks like a small portion of an active (Seyfert-like) nucleus. That supports the conclusion, from their ubiquitous pairing tendency across the active nuclei, that they have been ejected in opposite directions from this central point, which shows similar physical conditions. As explained in the introduction, starting in about 1948, it has become an article of firm belief that galaxies eject radio emitting material in opposite directions. The quasars often show radio emission, as well as the other attributes of matter in an energetic state, such as X-ray emission and excited optical emission lines. *The only possible conclusion from this observational evidence is that quasars are energized condensations of matter which have been recently ejected from active galaxy nuclei.*

We will see later, however, that it will be necessary to consider the quasar to be made of more recently created matter in order to account for its higher intrinsic redshift.

Terminology

It is interesting to recount how the current confusion between some Seyferts and quasars came about. When the luminosities of quasars were computed on the assumption that they were at their redshift distance, it turned out there was continuity with galaxies in that parameter as well as other properties. Maarten Schmidt decided that

$M_V = -23.0$ mag. was about bright enough for a galaxy; and that anything brighter than that should be called a quasar. Of course, it has turned out quasars are actually *fainter* than galaxies, and should be classified on the empirical criteria of compactness and spectral excitation.

Another example of the penalty people pay for not using operational definitions is the term "AGN" (active galaxy nuclei). Once, as I was stepping onto a plane in Santiago, heading back to Pasadena, I met Bruce Margon coming the other way for an observing run in Chile.

"Oh Chip", he enthused, "I have just decided the new terminology for all these objects: we'll call them AGN's." He was using his theoretical knowledge that quasars were enormously bright nuclei of enormously distant galaxies.

"Absolutely terrible", I replied, "If you do that you will wreck the empirical classification."

Eventually, everyone came to believe that quasars had host galaxies. John Hutchings, Susan Wyckoff, Peter Wehinger and others found host galaxies. Assuming the quasars were at their redshift distances, they found host galaxies that were too big— and some examples that were too small. Taking a mean, they reported that their sizes were just right. When the Space Telescope started taking high-resolution pictures of quasars, John Bahcall called a press conference to report that a number of them did not have any host galaxies at all! Gasp! *Naked quasars!!*

The community was horrified. What was going to sustain the enormous luminous output of distant quasars if they did not have a host galaxy to fuel them? Private meetings were held immediately, and it was rumored that incorrect image reduction was involved. The judgment of doom! The irony here was that Bahcall had been coming on like Ghengis Khan, Tammerlane and Vlad the Impaler to anyone who doubted the redshift distance of quasars. Bahcall then produced some quasars with "host" galaxies, and everyone decided to paper over the issue in public.

There was no need for this chaos because the first quasar discovered (3C48 by Matthews and Sandage; 3C273 was only the first to have its redshift determined) had a nebulous fuzz around it, about 12 arc sec in extent. At a conventional distance corresponding to its redshift of $z = .367$, this translated into a diameter of 35-70 kiloparsecs, depending on the choice of Hubble constant. That is bigger than the big galaxies we know the most about, *e.g.* M31 and M81. But many quasars with a z around .3 were observed to have central brightness 3 or more magnitudes fainter than the 16.2 mag. of 3C48. Observed with seeing better than 1 arc sec., many showed no fuzz at all, so any host galaxy would have had to be abnormally small. Figure 2-18 shows a long exposure of 3C48—not with Space Telescope, but with a relatively small aperture 2.2 meter telescope in Hawaii and some image processing. It shows that the quasar has slipped completely out of the alleged host galaxy! What a way to fuel a quasar! What is worse, anyone who bothered to look would see that a huge low surface-brightness envelope surrounds the pair. The galaxy looks very much like a nearby dwarf!

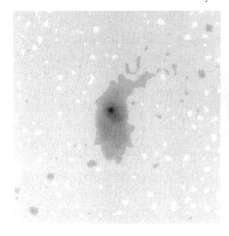

Fig. 2-18. The nearby quasar 3C48 as registered on co-added photographs with the Hawaiian 2.2 meter telescope by Allan Stockton *et al*. Note the quasar slipping out of the nucleus of the "host" in the blow up on the right. On the left there appears an extended, low surface brightness envelope around the system which looks like a dwarf galaxy.

If scientists had only heeded the words of Percy Bridgeman on the necessity of scrupulously using operational definitions in science. It would now be natural to describe an empirical sequence of quasar development, from initially point-like objects at relatively faint apparent magnitudes, gradually transforming into lower-redshift compact objects with "fuzz" around their perimeters, then into small, high surface-brightness galaxies with more material around them and, finally, normal, quiescent galaxies.

Trying to Stop the Stampede

When just the most prominent members of operationally defined classes are known, it is usually easier to see the overall relations between them. Figure 2-19 shows the Hubble diagram which I published in June 1968 (*Astrophysical Journal* 152,1101). The diagram showed that compact galaxies (morphological transitions between galaxies and point-like quasars) had active, Seyfert-like spectra and formed an obvious physical continuity between Seyfert galaxies and quasars. But, as Figure 2-19 shows, *this class of objects clearly violated the Hubble redshift-apparent magnitude relation.*

Nevertheless, this very Hubble relation is assumed in order to calculate luminosities for these objects. Then the luminosities are used to reclassify them on the basis of a theoretical property, which leads to the chaos described above. I followed the June paper with an expanded version in July 1968 (*Astrophysical Journal.* 153, L33) in which I showed more members of these classes which were continuous in color properties as well, and even more conspicuously violated the slope of the Hubble line. But my desperate effort did not even slow down the rush to express all measured quantities in terms of great distances in an expanding universe. The juggernaut has continued to gather momentum to the present day.

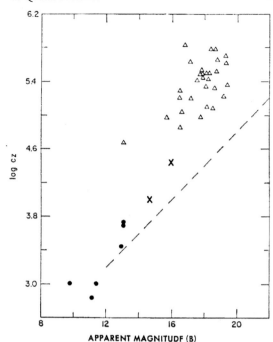

Fig. 2-19. The Hubble diagram for objects with Seyfert like spectra published by Arp in 1968. The solid circles represent nearby Seyfert galaxies, the crosses represent compact galaxies with Seyfert-like spectra and the triangles represent quasars known at the time. The class of objects obviously violates the dashed Hubble line which objects of the same luminosity in an expanding universe must obey.

3C48 as a Key to the Paradigm

We will show in Chapter 5 that 3C273 (the first quasar to have its redshift measured and, on the redshift-distance assumption, discontinuously the most luminous) is an important member of the relatively nearby Virgo Cluster. But the first quasar discovered was 3C48, and from it one could correctly deduce that it was a very strong radio source and a bright apparent-magnitude, stellar-appearing object. One might also suppose that of all members of this class of objects, it was among the nearest to us. Then, if the preceding Chapters have any meaning, one would expect a *very* bright, low-redshift galaxy to be identifiable as its progenitor at not too great a distance from it on the sky.

Now, one of the brighter galaxies in our Local Group of galaxies is M33, a companion to the dominant M31. M33 is a companion galaxy with a rather young stellar population, and just the kind of galaxy first associated with quasars (see *Quasars, Redshifts and Controversies*). The quasar 3C48 is only about 2.5 degrees away—exceptionally close for such bright objects! Figure 2-20 shows the configuration with another bright quasar in the region. If M33 were removed to the distance of the Virgo Cluster, the angular separation of 3C48 and paired quasars would be 7.1 and 12.9 arcmin from the galaxy. This is just the range of separations we were finding for quasars at the beginning of this Chapter, around galaxies which were on average at just about the distance of the Local Supercluster center.

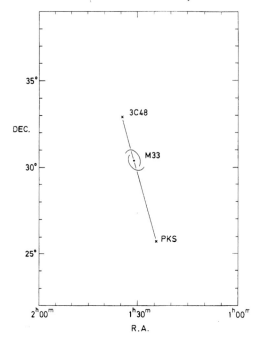

Fig. 2-20. This Figure illustrates the proximity of the first discovered quasar, 3C48, to the bright, Local group companion, M33. On the other side of M33 is the exceptionally bright, high redshift quasar, PKS 0123+25 with V = 17.5 mag. and z = 2.353 (see text).

What about the quasar on the other side of 3C48? Its redshift is z = 2.353, and it is a strong radio source with very bright apparent magnitude V = 17.5 for such a large redshift. (Further out in this region in this direction we see an extension of more high-redshift quasars, which apparently belong to M33, as shown in *Quasars, Redshifts and Controversies* pp. 72-73). But we also know from the just referenced work that the high-redshift quasars are less luminous than the lower-redshift quasars. This supports the surprising result that quasars of redshift up to about z = 1.5 can be seen out to the distance of the Virgo Cluster, but quasars of greater than about z = 1.8 are generally not seen much beyond the bounds of the Local Group.

Actually, the PKS quasar in Figure 2-20 is probably a secondary ejection. The candidate for the counter ejection from 3C48 would be a bright BL Lac (15.7 mag., redshift unknown) at 1h 09m 24s and 22d 28m 44" (1950). Because of the rapid evolution of high redshift quasars (z around 2 or greater), we would expect them to be seen rather close to their galaxy of origin. The latter prediction is forcefully born out by the 7 high-z quasars around the Seyfert 1 galaxy 3C120, which appears to be the closest active galaxy to our own in the Local Group. (See page 130 in *Quasars, Redshifts and Controversies*. That book also contains a chapter on the distribution of high-redshift quasars in space (Chapter 5), which shows their locations in the Local Group, with the strongest concentration southwest of M33 (lower right in Figure 2-20).

Way Back in the Beginning

In about 1951 I was choosing a Ph.D. thesis topic. I had been captivated by the early reports of Karl Seyfert's discovery of galaxies with brilliant compact cores. I was particularly intrigued with the fact that these cores were rich in ultraviolet light. I guess I sensed this was where there was some action, some mystery. For a thesis, I proposed to photograph these galaxies in ultraviolet light, and see what connection the nucleus had to the galaxy, and whether there were any other ultraviolet objects around.

Rudolf Minkowski, who was Walter Baade's right hand man, said that was a terrible thesis that would yield nothing. I wound up measuring thousands of little clumps of silver grains (photographic images of stars in globular clusters) in order to calibrate distance indicators in which Baade was vitally interested. Twenty years later, I was finding quasars around active galaxies by photographing them in ultraviolet and blue light and taking spectra of those candidates with ultraviolet excess. Occasionally, I would think on those nights: If I had done that thesis, maybe I would have discovered quasars ten years before they were discovered from radio positions. What difference would it have made to the course of cosmology? Then again, maybe I wouldn't—and then I would not have gotten the chance later.

Even though the globular cluster thesis-work helped lead to derivations of the age of the oldest stars, and hence to the age of our galaxy, Baade was suspicious of my reliability and did not recommend me for a staff position. It was Allan Sandage who successfully pressed for my appointment, because he thought I would be a great help in determining the Holy Grail of the distance scale that was the keystone of cosmology. But when I started having independent opinions about stellar population types that proved too competitive for Allan, and he wanted to get rid of me. When that did not happen, he refused to speak to me for ten years. Later he began to feel lonely, and we were close confidants for a while. One day he sat in my office and said, "Chip, you're the only one I can talk to." Well it was up and down a lot after that, too. But in the end, regardless of everything else, I feel close to him—like someone you have been together with through a tough war. It transcends the issues, and even the opposite sides, because no one else quite understands.

Chapter 3

EXCESS REDSHIFTS
ALL THE WAY DOWN

There is a story about a cosmologist giving a public lecture. Afterwards a lady stood up and said, "The real universe rests on the back of a turtle." He quickly shot back, "Well what is the turtle standing on?" "Young man," replied the lady, "it's turtles all the way down."

For those astronomers who are willing to consider quasars much closer than their redshift distances, there is one great stumbling block. That block is the instilled certainty that "normal" galaxies can only have velocity redshifts. When it comes to intrinsic redshifts in galaxies, most astronomers would consider that to be "turtles all the way down."

Yet we have already seen signs that quasars are not the only objects in the universe to have intrinsic redshifts. This would almost have to be the case just from considerations of continuity. There is a very obvious continuous progression of empirical characteristics from unresolved high-redshift quasars, through lower redshift compact objects, and finally into normal galaxies. We can argue that this is simply evolution in age, because the compact objects must be young—both from their tendency to expand due to the outward pressure of the concentrated energy, and the fact that the high energy tends to decay unless strongly fueled. Actually though, I was led to look for intrinsic redshifts in companions to large galaxies by an empirical series of results.

Companion Galaxies

The *Atlas of Peculiar Galaxies* contained a very interesting class of galaxies called spirals with companions (smaller galaxies) on the ends of arms. How had they got there? Certainly not by accidental collision or by the beginning of a merger process, which is

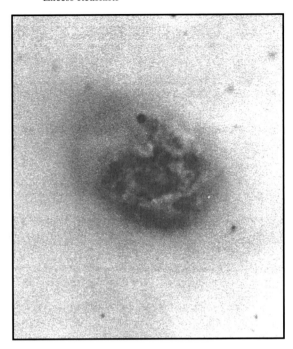

Fig. 3-1. No. 49 in *The Atlas of Peculiar Galaxies* shows a compact object trailing a wake of material behind it as it passes out through the disk of the galaxy.

fashionably used to "explain" everything in the extragalactic realm. (I actually read in the *Astrophysical Journal* once that double galaxies are galaxies in the process of merging and single galaxies are galaxies that have already merged.) I had argued that since galaxies characteristically eject material which eventually forms new galaxies and that if ejection took place in the galactic plane, then it would pull material out in the form of a spiral arm attached to the companion. Figure 3-1 here is No. 49 in my *Atlas of Peculiar Galaxies*, and it suggests quite plainly what is going on.

Whether or not that is true, I decided to look at the redshifts of the companions to see if, by any chance, they were systematically greater than the larger galaxy. They were, and that started another long running battle which eventually led to a quantitative proof of the dependence of redshift on age.

The clues begin in the Local Group of galaxies centered on our giant Sb spiral M31, historically known as "the Andromeda Nebula." M31 is the most massive galaxy in our group, and is classified Sb by virtue of its extensive central bulge of old, red stars. Every major companion (by inference, including our own Milky Way galaxy) is positively redshifted, as seen from M31. The next nearest major group to us, the M81 group, is centered on the same kind of massive Sb galaxy and, again, every major companion is redshifted with respect to it!

By 1987, there had been a dozen different investigations, every one of which showed companion galaxies were systematically redshifted (see Table 7-1 of *Quasars, Redshifts and Controversies*). By 1992, there were 18 different references to studies which showed this effect in the published literature. In spite of all this, a paper then appeared in the *Astrophysical Journal* which interpreted companion redshifts as velocities to derive

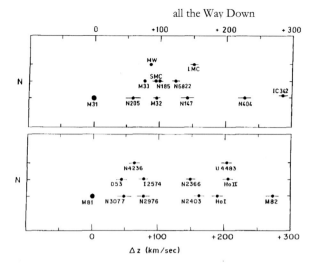

Fig. 3-2. The Local Group (M31) and the next nearest major group (M81). The smaller, companion galaxies are shown to be all of higher redshift, a distribution having one chance in 4 million of being accidental.

masses of parent galaxies—and referenced none of the 18 papers which showed that the velocity assumption was untenable. I got pretty heated up over this, and after a lengthy battle managed to get an answering paper published in the same journal (*Astrophysical Journal* 430,74,1994). Figure 3-2 here is taken from that paper.

One interesting development that had taken place was that a new member of the Local Group of galaxies had been found, IC342. This dwarfish spiral was at low galactic latitude (Figure 3-3), and an accurate absorption and distance had only been determined recently. It then became a member of the Local Group at about 1.2 Mpc distance on the other side of M31 from us. At +289 km/sec redshift with respect to M31, it had the largest excess redshift. (Actually this redshift was very close to four times the basic redshift quantization of 72.4 km/sec, a matter that will be discussed further on.) This discovery brought the count to 22 out of 22 of the major companions, all of which had higher redshifts than the dominant galaxy, in the two best known, nearest groups. *The*

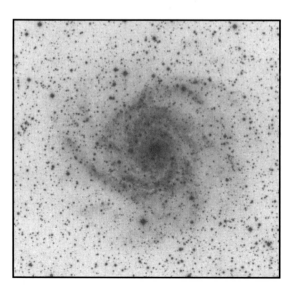

Fig. 3-3. A spiral of large apparent diameter seen close to the plane of the Milky Way. IC342 is the newest and most distant member of the Local Group.

chance of this arrangement of galaxies randomly orbiting their central galaxies with equal numbers of approaching and receding velocities was only one in 4 million!

Major Clusters of Galaxies

If companion galaxies in groups have systematically larger redshifts, what about companions (less luminous) galaxies in clusters? One could logically argue that great clusters, like the Local Supercluster, were made up of many groups like the M31 and M81 groups. In fact, it is true that if one looks at the Virgo Cluster (*i.e.* the center of the Local Supercluster), one finds all the usual morphological types of galaxies. And the smaller galaxies are systematically redshifted with respect to the larger!

One can see this effect in two ways. First one can calculate the mean redshift of the galaxies in the Virgo Cluster by weighting the redshift of each galaxy by the brightness of the galaxy. If luminosity is proportional to mass, then one gets the redshift of the average mass of the cluster, the only dynamically meaningful quantity. This calculation gives a mean redshift for the Virgo Cluster of +863 km/sec. Now the value calculated by assuming all the galaxies have the same mass comes out to between 1000 and 1200 km/sec. *Why this striking difference? It is simply because the smaller galaxies have systematically higher redshifts.*

The second way to see this effect is to note that late-type galaxies (spirals and young spirals) are systematically redshifted in clusters. Since spirals are generally less luminous than giant E's, and, further, since their mass-to-luminosity ratios are lower; this shows in a different way that companion (lower-mass) galaxies in clusters are systematically redshifted.

The Redshift of the Virgo Cluster and the Hubble Constant

Sometimes I think that Astronomy is not so much a science as a series of scandals. One of the most egregious is the derivation of the value of the Hubble constant from the Virgo Cluster. There have been innumerable headlines about new distance determinations to the cluster in the past decades, and most recently from Space Telescope press releases. The debate swings between the "long" distance scale (a little more than 20 megaparsecs) and the "short" distance (about 16-17 Mpc). The longer distance is used by the proponents of $H_o = 50$ km/sec/Mpc. The shorter distance is used by proponents of H_o around 80, the latter having the drastic consequence that the universe is then younger than the oldest stars it contains. (Unless one brings back the cosmological constant *etc., etc.*)

Although both sides use different mean redshifts for Virgo (ones that favor their preferred value: see *Astronomy and Astrophysics* 202,70,1988), neither side pays the slightest bit of attention to the fact that they have both made an elementary mistake in computing that mean. In physics, we learn to compute the center of mass of an ensemble of particles by weighting each particle. How can we compute the mean redshift of the center of mass of a cluster of galaxies without weighting the mass of the galaxies? Of course, astronomers insist on assuming that the low-mass and high-mass galaxies have the same average redshift. If that were so, they would get their usual answer, and they should have no objection to making the more rigorous calculation. In fact, if they defined the dynamical center as the most luminous and massive galaxies (which should not drift away from the rest of the cluster), they would not be able to change the mean redshift of the cluster by adding or not adding negligibly small galaxies over which there is obvious disagreement as to membership.

Another "adjustment" which pushes the derived Hubble constant to higher values is the notion that the mass of the Virgo Cluster attracts our own Local Group, and its consequent "infall velocity" must be added to obtain the true cosmic recession velocity of the Virgo Cluster. The "infall velocity" is the supposed result of the gravitational attraction of the Virgo Cluster on the Local Group. But if masses of galaxies have been generally overestimated, or if peculiar velocities between groups are very small—both of which will be argued later—, then this adjustment cannot be used to increase the Hubble constant, as in the conventional derivation. Moreover, if galaxies on the near side of the Virgo Cluster were falling toward its center, then the brightest galaxies would have the more positive redshifts. The opposite is actually true. Therefore, the 1400 km/sec systemic redshift used for the much-publicized Hubble constant calculations is far from the 863 km/sec actually measured. (In fact, 863 km/sec is an overestimate because luminosities measured in red wavelengths should be used and also the spirals weighted less.)

Late type Spiral Galaxies as Younger Companions

From the beginning, we have noticed the excess redshift of companions around massive central galaxies which had large components of old stars. The implication was that these old stars had been around from the beginning of the group, and that smaller, younger companions had been ejected intermittently as time passed. These central galaxies had morphological types mainly of Sa, Sb and giant E. The smaller companions ranged over the remaining morphological types, but featured dwarf E's (showing spectroscopic indications of an admixture of a population of stars younger than in the giant E's) and later-type spirals (SBbc, Sc, Sd and Im). The latter types are marked by conspicuous numbers of bright, young stars. These late-type spirals were measured to have low masses from their rotation characteristics, and low mass-to-luminosity ratios indicative of relatively recently formed stars. Empirically then, the smaller nuclear bulges and open spiral structure of the late-type spirals came to mark them as lower-mass, younger "companion" type galaxies.

A special kind of supposedly high luminosity spiral, designated ScI, will be discussed later as really being of low luminosity because of large excess redshift due to its younger age. But for the purpose here of investigating the redshift behavior of companions in major clusters of galaxies, it will be useful to identify companions by their morphological classification as late type spirals.

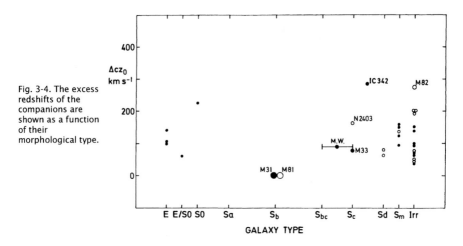

Fig. 3-4. The excess redshifts of the companions are shown as a function of their morphological type.

Late Type Spirals in Major Galaxy Clusters

Figure 3-4 shows the excess redshift of companions as a function of their morphological type in the two nearest groups, M31 and M81. The later type spirals are clearly systematically higher redshift. Figure 3-5 shows the same diagram for the entire Virgo Cluster, and the same pattern is evident. This enables us to check other major clusters, as shown in Figures 3-6 and 7. The end result is that the younger spirals in the nearest groups, as well as the 4 or 5 major clusters of galaxies, *all* show systematic positive redshifts. There seems to be no escape from this result.

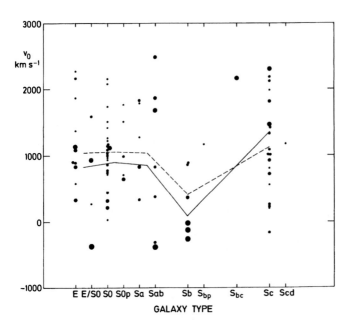

Fig. 3-5. Redshifts of galaxies in the Virgo Cluster as a function of morphological type. The full line is the luminosity weighted mean and the dashed line the number mean. Symbol sizes are proportional to the apparent magnitudes. Note that, as in the nearby groups, the galaxies around type Sb are the lowest redshift and tend to be among the brightest.

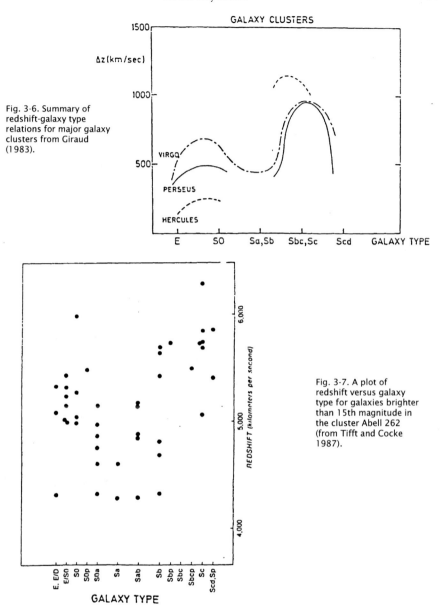

Fig. 3-6. Summary of
redshift-galaxy type
relations for major galaxy
clusters from Giraud
(1983).

Fig. 3-7. A plot of
redshift versus galaxy
type for galaxies brighter
than 15th magnitude in
the cluster Abell 262
(from Tifft and Cocke
1987).

Back to the Virgo Cluster

In Figures 3-4 and 3-5, what is really most apparent is the minimum redshift
exhibited by the brightest Sb's. In the Virgo cluster, galaxies of this morphological type
are predominantly low or even *negative* redshift. One could obtain a very low redshift for
the cluster if one were to accept them as the dominant galaxies in the cluster.

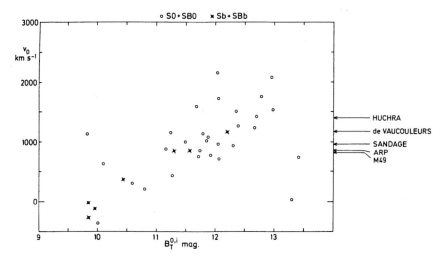

Fig. 3-8. Redshift-apparent magnitude diagram for all of the S0 (open circles) and Sb (filled circles) galaxies which are the most certain members of the Virgo Cluster. The mean redshift for the cluster by various authors is indicated along the right ordinate. The Huchra value includes an infall velocity.

The S0's (a kind of disk galaxy without bright young stars), which are the most numerous kind of galaxy in the cluster, actually show a continuous gradient of redshift from the brightest to the faintest. Figure 3-8 shows their apparent magnitude-redshift relation. Again one could pick almost any redshift for the Virgo Cluster one wanted, depending on the apparent magnitude of the S0 which one chose to be representative of the mean mass. *On the conventional assumption, the S0's are supposed to define a horizontal line in Figure 3-8!* In view of this uncertainty, the best procedure seems to be to make a luminosity weighted integration over the galaxies in the cluster and hope that this averages close to the age of our own Milky Way galaxy, so that there is no age induced differential redshift.

It is encouraging to note that the +863 km/sec derived in this way is very close to the largest, brightest and apparently oldest galaxy at the geometrical center of the Virgo Cluster, M49 (also known as NGC4472). That seems the best bet to be the currently dominant galaxy and it has a redshift of +822 km/sec. If we take the short distance scale to the Virgo Cluster of 16-17 Mpc (in my opinion the more correct one) we obtain a Hubble constant, H_0, close to 50 km/sec/Mpc. We will see in Chapter 9 that this fits quantitatively with a non-expanding universe in which the redshift is a measure of the age of a galaxy.

What about the negative redshifts in Virgo (*i.e.* blueshifts)? People often ask: If intrinsic redshifts are a function of age, can there be negative redshifts? The answer is: Yes, it is required if the galaxy is older than we are, as we see it. Aside from the Local Group where M31 is the parent and we see it as negatively redshifted by −86 km/sec, there are only six major galaxies of negative redshift in the sky. All six are in the Virgo Cluster, and are obviously members. They are chiefly the big Sa's and Sb's which we

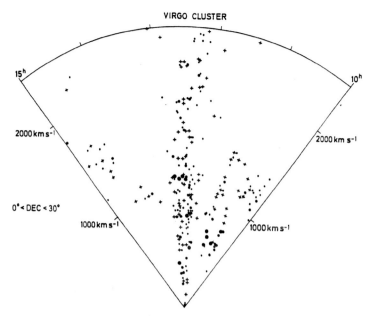

Fig. 3-9. A pie diagram for all galaxies listed as Virgo Cluster members in the Revised Shapley Ames Catalog by Sandage and Tamman, plotted as a function of their redshifts. Crosses are spirals and later types, pluses are remaining types. Symbol size decreases with decreasing apparent brightness.

have learned to associate with the originally dominant galaxies. Hence, these are probably somewhat older than any of the galaxies in our Local Group, and may represent the original galaxies in Virgo. It is even possible that our Local group originated from them. [It is touching to speculate that when we look at the Andromeda galaxy, we are looking at our parent. Perhaps in Virgo we can gaze at our grandparents.]

Later, we will discuss aggregates of numerous faint smudges which are called distant galaxy clusters. But we will argue that they are generally something different from the great clusters of galaxies like to our own.

Pie in the Sky Diagrams

An enormous amount of modern telescope time and staff is devoted to measuring redshifts of faint smudges on the sky. It is called "probing the universe." So much time is consumed, in fact, that there is no time at all available to investigate the many crucial objects which disprove the assumption that redshift measures distance. Still, one has to do something with these redshifts after they are measured. What is done is, an area on the sky is selected and all the available measures plotted as a function of their redshift.

As an example, the plot in Figure 3-9 shows what the well-known Virgo Cluster looks like. What a shock! There is a great "Finger of God" pointing directly at us, the observer. Of course, this is hastily explained as due to high orbital velocities for the galaxies in the center of the cluster which invalidate their use as distance criteria. But it is not just the center of the cluster which shows these "peculiar" velocities; the whole

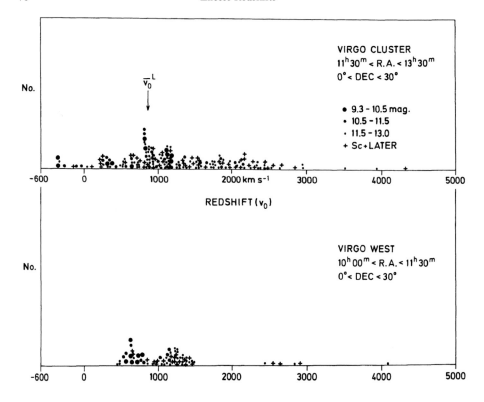

Fig. 3-10. The same galaxies as in the preceding diagram but now plotted as a distribution function of redshifts which enables negative redshifts to be included. The luminosity weighted mean is indicated by an arrow.

cluster is strung out. Moreover—and this is the telling point—, the brightest galaxies are preferentially at the lowest redshift. This is shown even more clearly in Figure 3-10, where the negative redshifts in Virgo can be plotted. The fainter galaxies and late type spirals trail asymmetrically away to much higher redshifts. *If the elementary precaution of plotting these points in proportion to their brightness had been taken, it would have been obvious that the fainter galaxies had intrinsic redshifts.*

Another obvious feature of Figure 3-9 is that the higher-redshift tail drifts off in a different direction from the center of the Virgo Cluster. That cannot be due to velocity dispersion in the center of the cluster. These must be smaller galaxies in a somewhat different part of the cluster, but with a continuity of increasing intrinsic redshift. This one feature, by itself, is disproof of the redshift-equals-velocity hypothesis.

Nevertheless, region after region of the sky has been presented in journal articles and public lectures that present the Fingers of God as velocity dispersions and show how the universe is made up of bubbles and voids. When people occasionally question this orgy of Swiss cheese universes, the answer is always the same: anyone who does not believe redshifts are measures of distance is termed a "psychoceramic artifact."

SPHERICAL GALAXY CLUSTER

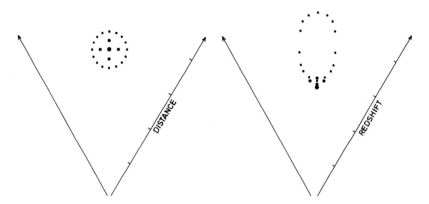

Fig. 3-11. Illustration of what happens when one takes a spherical galaxy cluster with the brightest, lowest redshift galaxies in the center and then plots them in a pie diagram with redshift assumed to be a measure of distance.

Blowing Bubbles and Digging Voids

Considering what we know about a group or cluster of galaxies, let's look for a moment at how plotting them in a pie diagram would distort the picture. Figure 3-11 shows the large, low redshift galaxies at the center with the smaller, intrinsically redshifted galaxies distributed around them. As soon as we plot with redshift as a distance indicator, the large low-redshift galaxies pull out of the center, leaving a ring or bubble.

There are many ways one could elaborate on this picture. If the fainter galaxies have plunging orbits, that would elongate the ring along the line of sight. If there is a component of rotational orbiting, that would fatten the ring toward the edges. An approximate mixed velocity dispersion is shown in the right hand panel of Figure 3-11. Of course, this is all under the usual default assumption that the cluster or group is in equilibrium. One could find a variety of forms if groups of younger galaxies were moving away from the central galaxy.

Since we know that the central, larger galaxies have the lowest intrinsic redshifts, it will now be necessary to go back and carefully correct the inferred distributions of galaxies in different directions in the sky.

Further Evidence for Excess Redshifts of Companion Galaxies

Shortly after the publication of my 1994 paper on the subject of companions described earlier, I was walking past the journal rack in the library when a paper on "Arp 105" caught my eye. Curious, I skimmed it and quickly ascertained that, as with so many other objects from my *Atlas of Peculiar Galaxies* which were prime examples of

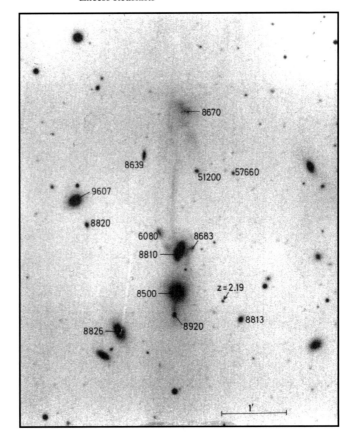

Fig. 3-12. Photograph of Arp 105 (NGC3561B). Ambarzumian's knot is seen ejected due south from this active elliptical and an opposite ejection northward appears to be puncturing the disturbed spiral. Redshifts of galaxies measured by Duc and Mirabel and quasar of redshift $z = 2.19$ discovered by Alan Stockton are indicated.

ejection, this was also being presented as an example of collision and merger. As Figure 3-12 and the color presentation on the back cover of this book testifies, this was a particularly inappropriate interpretation, because it was one of Ambarzumian's finest examples of protogalaxies being ejected, jet-like, from an active elliptical galaxy. Exactly opposite this was the counter jet, a magnificent straight plume punching through a disrupted spiral. Fritz Zwicky, after looking at his spectra of the knots in the jet, had remarked that these were the only galaxies he knew that were not resolved with the 200-inch telescope. Allan Stockton had discovered a quasar of redshift $z = 2.2$ so close to this ejecting galaxy that the chance of accidental occurrence was less than one in a thousand.

I was about to return the paper to the stand with exasperation when I noticed that the authors had measured the redshifts of most of the companions. What they had overlooked, and what leaped off the page, was that they were all positively redshifted with respect to the dominant galaxy. Since the authors claimed these companions were colliding with what they termed a "giant E", there was no question that they believed the galaxies they had measured were *bona fide* companions at the same distance as Arp 105. It did not matter whether they were orbiting the central galaxy, falling in, or being ejected outward—one should roughly expect just as many relatively plus as minus

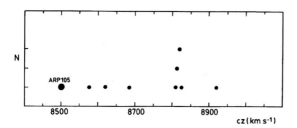

Fig. 3-13. Distribution of redshifts of companion galaxies around the "massive elliptical" Arp 105 as measured by Duc and Mirabel.

velocities on the average. As Figure 3-13 reveals, all 9 of the lowest redshifts (actually 10 if one were to include the one at +1100 km/sec, relative redshift) are higher than the central galaxy redshift. One had here another case, like the Local and M81 groups, where the intrinsic excess redshift of the companion had overcome the smaller plus and minus velocity dispersion.

One of the reasons this was a particularly satisfying confirmation was that this was a somewhat different kind of central galaxy, much rarer, caught in the act of ejecting. It had a much higher mean redshift than the more local groups that had been tested. In addition, there was an unusually large number of companions.

While I was writing this result up for communication, a preprint crossed my desk. A new investigation of the Hercules Cluster of galaxies had shown that in every subsection of the cluster, the late-type spirals (companions) had conspicuously higher redshifts than the early-type galaxies in the same sector. This was impressive, because it was a *detailed* confirmation of the results for companions in clusters.

Finally, simultaneously with the above, a student in Holland sent me one of the secondary findings in his thesis. While investigating galaxies in the Bootes void, he had discovered that 78% of the companions around his galaxies had positive redshifts relative to the dominant galaxy. Figure 3-14 shows this very strong confirmation in a large sample of galaxies.

Fig. 3-14. Excess redshift for companions in a sample of galaxies in the Bootes void and comparison fields as measured by Arpad Smozuru. Plotted as a function of the distance from the central galaxy.

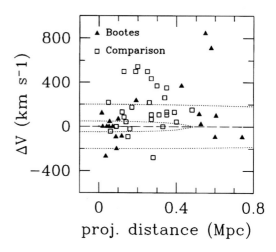

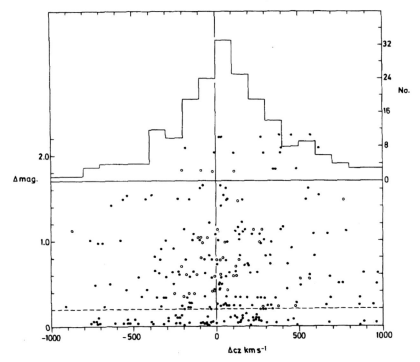

Fig. 3-15. For Paul Hickson's catalog of compact groups of galaxies, the amount by which
the brightest galaxy is brighter than the second brightest is designated Δmag. The
redshift of each galaxy in the group minus the redshift of the brightest in the group is
called Δcz. The histogram shows that for mag. > .2 mag, the fainter companions are
systematically redshifted.

Trying to Publish Further Results

Putting all these results together, they seemed to me to offer decisive proof of
excess redshifts in companions. But the author who used the companion redshifts as
velocities, without referencing the contrary evidence, wrote an angry letter to the editor
complaining that I had been rude in my manner of pointing this out. Another pair
published a rebuttal paper claiming complex orbits could explain the preponderance of
positive companion redshifts! When the "Further Evidence" paper went to the referee,
he suggested the interacting companions around Arp 105 belonged to another galaxy
outside the pictured area. There were hints that the thesis student who found the excess
companion redshifts would be in big trouble. After holding the paper for three months,
one referee sent a Xerox from a 1902 book on celestial mechanics plus a graph showing
the moon orbiting around its barycentre. Another referee said a study of weak galaxy
clusters showed the largest galaxies to have the same redshift as their cluster. When I
analyzed that data, the same result turned up—the brightest galaxies had −355 km/sec
lower redshift. The referee replied to the editor: "Perhaps the author did not understand
that I have rejected the paper."! The editor rejected it.

At a conference, one of these referees gave a rather startling (for conventional beliefs) lecture on how Fourier analysis could not be trusted, and mentioned that ergodicity did not ensure that the ensemble average was equal to the time average! He said he was eager for more data on this important subject of companions!! But after *four years*, this further evidence had not been published in a major journal. The only result is a stack of insulting letters from referees and editors.

One thing has been accomplished, though. I now understand what should be called the statistics of nihilism. It can be reduced to a very simple axiom: "No matter how many times something new has been observed, it cannot be believed until it has been observed again." I have also reduced my attitude toward this form of statistics to an axiom: "No matter how bad a thing you say about it, it is not bad enough."

Compact Groups of Galaxies

The first compact group was discovered with the Marseille telescope in 1877 by M. E. Stephens. In 1961, Margaret and Geoffrey Burbidge measured redshifts of the five galaxies and showed they were 800, 5700, and three at 6700 km/sec (see Figure 3-19). Now the 5700 and one of the 6700 galaxies were entwined together. If redshifts were interpreted as velocity, this meant they were separating at 1000 km/sec. Even in conventional terms, galaxies don't move that fast; and even if they did, the chance of catching two at just the moment of collision would be very small. And, of course, the gas would not keep two separate velocities. From that time forward it should have been clear we were dealing with non-velocity redshifts.

But as you might suppose, the picture has become increasingly muddied with mergers, dark matter and gravitational lenses, while any redshifts which do not fit a conventional theory are placed in the foreground or background. Is there anything new? Well, an observational advance has been made by Paul Hickson, who catalogued, photographed and measured redshifts in a sample of 100 compact groups. (A compact group is defined as four or more galaxies crowded together by a factor of 10-30 more than their local surroundings.) The *Catalogue* made it possible to test the following proposition: Since compact groups are in many cases denser versions of normal groups in which companions have excess redshifts, do compact groups with a dominant galaxy have systematically redshifted companions? Figure 3-15 answers this question by showing that, as the difference in apparent magnitude between the brightest and next brightest galaxy becomes larger, the number of positively redshifted companions becomes larger. This makes sense, because if the galaxies are all the same brightness, one does not know which is dominant and the effect is untestable. *But the fact that when one galaxy becomes clearly dominant the effect emerges—this demonstrates that non-velocity effects are present in the compact group galaxies, just as in every other group tested.*

This point is strikingly illustrated in Figure 3-16 where the distribution of companion redshifts in compact groups with the most dominant galaxies is compared to the Local Group. Actually most groups have companions with up to about 800 km/sec higher redshift, and it is obvious that the Local Group is missing companions

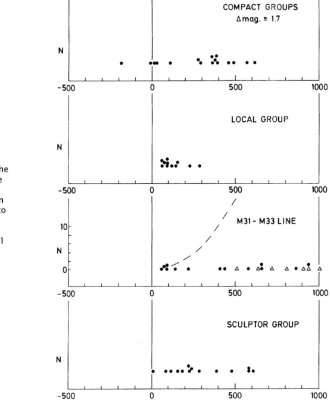

Fig. 3-16. This figure compares the excess redshifts of companion galaxies in the compact groups versus those of the accepted members in the Local (M31) group, then the fainter companions in the M31 line and finally to the companions in the small Sculptor group which is between the M31 and M81 group.

above about 300 km/sec. The reason is quite simple—namely that astronomers are just unwilling to call any galaxy more than 300 km/sec higher than M31 a member of the Local Group because that makes the preponderance of positive redshifts embarrassingly obvious.

If one examines the brightest galaxies as they fall on the sky, however, it is immediately apparent that there is a loose string of them running out of M31, through M33 and ending close to 3C120 near the disk of our galaxy. (See Figure 8-1 in a later chapter). These galaxies have redshifts up to 900 km/sec and are obviously members of the Local Group. A group of later-type spirals called the Sculptor Group is located closer to us than the M81 group. As the last panel in Figure 3-16 shows, it also has higher redshift companions. (Details are available in *Quasars, Redshifts and Controversies* page 131 and *Journal of Astrophysics and Astronomy.* (India) 1987, 8, 241.)

An earlier study of what I had called "multiply interacting galaxies" comprised the most striking examples of what later came to be called the "compact groups." What I pointed out in that original study was that these multiply interacting groups preferentially occurred near large, low redshift galaxies. In some cases, for example

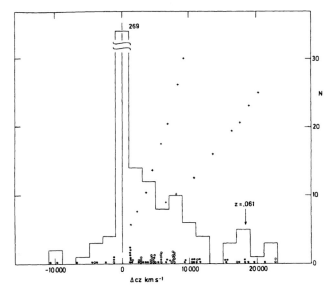

Fig. 3-17. Number of discordant (Δcz > 1000 m/sec) redshifts as a function of Δcz for compact groups. Lines of plus signs show expected distribution for background interlopers. Arrow points to preferred redshift peak of Δz=.061 found in all sky measures of quasar and quasar-like objects.

NGC3718, the high redshift, interacting group could be seen actually bending back the spiral arms of the larger galaxy. (Consult the picture on page 94 of *Quasars Redshifts and Controversies*.) This result made it clear that the compact and interacting groups were just a more concentrated ensemble of young, non-equilibrium companion galaxies which had been ejected more recently from the parent galaxy, and were composed of material of higher redshift. Aside from being empirically true, this interpretation solves all the conventional paradoxes of the failure of the galaxies to merge into a single galaxy on a cosmic time scale, and also explains the unbearable presence of "discordant" redshifts. Of course, none of this is conceded by the conventional army.

Large Excess Redshifts in Compact Groups

We have just seen that the so-called accordant group members (defined as having redshifts different from the group by less than 1000 km/sec) demonstrate again that the fainter members have the higher redshifts. But most shocking of all, there are a number of (mostly) fainter galaxies that fall in these compact groups which have redshifts thousands and tens of thousands of km/sec greater than the group (Figure 3-17).

The consternation caused by the apparent membership of these highly discordant galaxies has led to a blizzard of papers arguing that, despite appearances, they were just projected background galaxies. Just in case, it was also argued that they were gravitationally lensed background objects. To be triply safe, it was also argued that we were seeing filaments of galaxies end on—like looking down a straw with a galaxy stuck on the far end. The only trouble is that in the famous case of Seyfert's Sextet, the length of the straw had to be about 26,000 times its diameter (see *Astrophysical Journal.* 474, 74, 1997)!

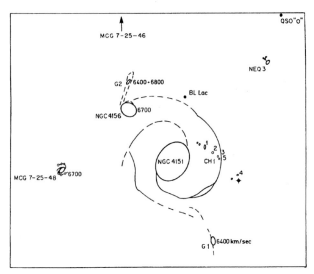

Fig. 3-18. Schematic representation of features of interest around the active Seyfert NGC4151. Note especially the companion galaxies at 6400 and 6700 km/sec.

Figure 3-17, however, shows with small plus signs how background galaxies should increase sharply in number with increasing redshift. The numbers of discordant redshifts actually decreases precipitously in this direction. It seems to me that at a glance, it is clear the discordants are not background galaxies.

Companion Galaxies of High Redshift

This is a very important link in the argument that material is ejected from large galaxies, initially with very high intrinsic redshift, and then ages and expands into compact, active galaxies of moderately high redshift, and finally into normal companions with only slightly excess redshift. So far we have shown that the extensive evidence which already existed has been enormously strengthened by new evidence that "normal companions" belonging to dominant galaxies have excess redshifts in the hundreds of km/sec. Companions with excess redshifts of thousands to tens of thousands of km/sec establish a compelling continuity to the quasars which start at about 20,000 km/sec excess redshift and go up to nearly the velocity of light (if they were really velocities).

Unfortunately, there is not much in the way of new results on this group. In 1982, a list of 38 (yes, thirty-eight) of these high redshift discordant companions was published. They were discussed in two *Astrophysical Journal.* papers and in a chapter starting on page 81 of *Quasars, Redshifts and Controversies.* Yet despite the fact that almost every one of these objects is a fascinating study in itself, no further study of these key objects has been made! *Certainly these crucial, discordant redshift galaxies have been deliberately avoided by the world's biggest and most expensive modern telescopes.*

To mention just two examples in order to reemphasize the importance of these kinds of companions, I show in Figure 3-18 a schematic of the large, active Sb, NGC4151. (Deep photographs of this galaxy can be seen in *Quasars, Redshifts and*

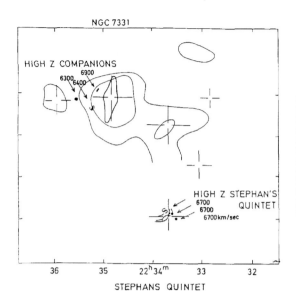

Fig. 3-19. Region around the large Sb, NGC7331 and Stephan's Quintet. Line contours represent radio emission. Note especially the companion galaxies at 6300 to 6900 km/sec around NGC7331 and the 6700 km/sec companions around NGC7320, the low redshift galaxy in the Quintet.

Controversies, pp. 91 and 92). We saw in Chapter 2 how this Seyfert was flanked by two pairs of quasar candidates in an apparent ejection cone, with a strong X-ray BL Lac inside this cone. But also around NGC4151 are associated large companions with redshifts between 6400 and 6800 km/sec. Two of these companions, NGC4156 and G1, lie at either end of the two major spiral arms. With their similar redshifts, are they not like the pairs of quasars discussed in Chapters 1 and 2? With the material of the arms trailing behind them, are they not reminiscent of ejection in the plane, as conjectured at the beginning of this chapter?

But perhaps equally striking is the *numerical value* of the excess redshifts of these major companions to NGC4151. If one refers to Figure 3-19, one sees that three galaxies in Stephan's Quintet also have 6700 km/sec redshift and the three galaxies roughly on the other side of the large Sb galaxy, NGC7331, have 6300, 6400 and 6900 km/sec redshift. Galaxies come in groups, and there is no other group leader for these ~6700 km/sec companions to belong to other than the big galaxies at their center. In later chapters we will show that galaxies and quasars tend to occur at certain preferred redshifts. This quantization implies that galaxies do not evolve with smoothly decreasing redshifts, but change in steps.

ScI Spirals as Young, Low Luminosity Galaxies

The companion galaxy NE of NGC4151 has the sharply defined spiral arms which define it as an Sc spiral of luminosity class I. This highest luminosity class is assigned because these galaxies characteristically have moderately high redshifts, which are taken to indicate large distances and high luminosities. Since it is attached to the low-redshift (978 km/sec) NGC4151, it in fact must have an intrinsic redshift and a low luminosity. The same is true of NGC7319, a high redshift ScI galaxy in Stephan's Quintet, which must be at the distance of the low redshift NGC7331 (1114 km/sec) (see Figure 3-19).

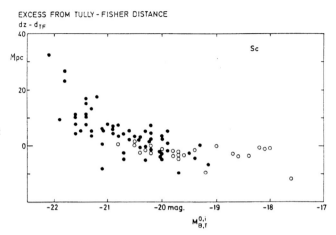

EXCESS FROM TULLY-FISHER DISTANCE

Fig. 3-20. The excess of
the redshift distance over
the Tully-Fisher distance
plotted as a function of
the luminosity of Sc
galaxies. Filled circles
represent
redshifts > 1000 km/sec.
This graph demonstrates
that for the highest
redshift, supposedly most
luminous Sc's, the
redshifts give distances
too large by huge
amounts.

How can we check this result? There is a method of estimating distances to galaxies, called the Tully-Fisher method, which uses the rotation of a galaxy to judge its mass, and thus its luminosity, and then its distance by how faint it appears. In Figure 3-20 we see the difference between the redshift distance and the Tully-Fisher distance plotted as a function of the supposed luminosity of the galaxy. We see that for normal spirals, the two methods are calibrated to give the same distance. But for the high-luminosity spirals (ScI's), the redshift distance is too great by up to almost 40 Mpc! This huge error demonstrates that the redshifts of the ScI's are too high.

A vivid illustration of how wrong astronomers' estimates of the sizes of ScI galaxies are is shown in Figure 3-21. The large galaxy is the ScI spiral NGC309 at its supposed redshift distance. The small oval insert shows the size of one of the largest galaxies for which we know an accurate distance, M81. The giant M81 is swallowed like a knot in the arm of the unbelievably large ScI. This picture was published in the April

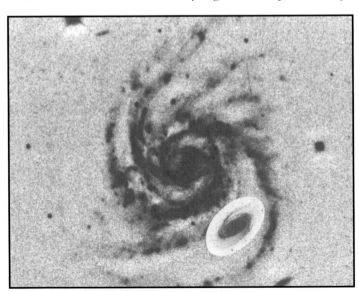

Fig. 3-21. The Sc, luminosity class I, NGC309, if it were at its conventional redshift distance would be so huge that it would swallow one of the largest galaxies of which we have certain knowledge, the Sb M81 (shown as an insert in the lower right between the arms of NGC309).

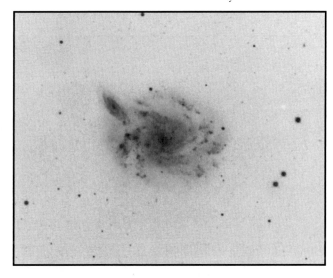

Fig. 3-22. The Sc galaxy NGC450 has a redshift of 1,900 km/sec and the smaller galaxy to the NE which is apparently interacting with it has a redshift of 11,600 km/sec. The three HII regions in NGC450 near the point of contact with the high redshift galaxy are unprecedentedly luminous and could only reasonably be explained by interaction.

1991 issue of *Sky and Telescope*, and the professional astronomers who saw it gasped in astonishment.

But the paper with the analysis was thrown out of the *Astronomical Journal* with great prejudice. When published in *Astrophysics and Space Science* 167, 183 it detailed a number of other cases where ScI's could be shown to be low luminosity, intrinsically redshifted galaxies. I speculate that the sharp, well-formed arms are young ejections before they have had time to be deformed and to spread out. But most astronomers are willing to suppress this observational evidence in order to protect the key assumption about extragalactic redshift from re-examination.

The "Non Interacting" Companion to NGC450

One case that was further investigated is the peculiar Sc spiral NGC450, shown in Figure 3-22. It has a redshift of 1,900 km/sec, and the apparently interacting companion has a redshift of 11,600 km/sec. Just at the point of interaction there appear three enormous HII regions at the redshift of the Sc galaxy. These were so gross that the expert who first spotted this system just assumed they were foreground stars. These glowing regions of excited hydrogen gas are so exceptional that I frankly cannot see how anyone with reasonable common sense and good judgment would not immediately realize that they are a result of the unusually close interaction with the companion.

Nevertheless, a pair of astronomers measured some rotation curves in the system, pronounced them "normal," and published a paper proclaiming "Non-Interacting" in the title. There would be nothing new to report if it was not for the Spanish astronomer Mariano Moles, who had long been intrigued with this system, and unknown to me, had conducted an extremely thorough observational project of photometry, spectroscopy and imaging on it with the moderate aperture telescope at Calar Alto. His analysis demonstrated six different observational results, all of which led to the conclusion: " ...

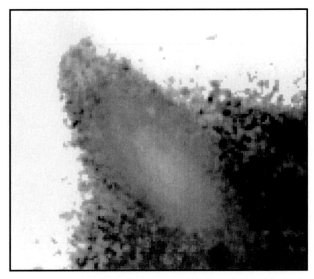

Fig. 3-23. The HII region which is at the SW end of the high redshift companion galaxy has a redshift that indicates it is falling from NGC450 into the companion. This picture shows a short luminous tail, supporting that interpretation.

one would have to invoke an enormous conspiracy of accidents in order to avoid the conclusion that [the companion] is a moderately low luminosity galaxy interacting with NGC450."

One particular aspect was especially pleasing to me. It involved the circumstance that on one of my last runs on the 200-inch telescope at Palomar, I had measured the redshifts of the bright HII regions on the companion side of NGC450. In particular, I had gone after the fourth and faintest HII region, which was just at the end of the high redshift companion, where the companion spread out in an apparent interaction effect with the lower redshift galaxy. It was a difficult observation, and I had to use the Oke multichannel spectrophotometer (commonly called the gold Cadillac). But the emission lines were strong and I got good measures, which I reduced before leaving California for Europe. The redshifts showed larger than normal differences of about 100 km/sec, but the faintest, near the end of the companion, showed a plus redshift of 400 km/sec, well in excess of escape velocity from NGC450.

That 400 km/sec measure enabled a very satisfying model of the interaction to be constructed. It was simply that NGC450 was rotating clockwise, and the companion was coming up from behind it. As the spiral arm of NGC450 approached the companion, its gas was being pushed back and was accumulating and forming the very large HII regions. The companion, which is wedging itself in between the spiral arms of NGC450, was close enough so that the nearest HII region was actually beginning to fall into the near end of the companion. The unexpected confirmation of this came from the hydrogen emission image, which showed a trail of excited gas as this fourth HII region fell toward the high-redshift companion (See Figure 3-23).

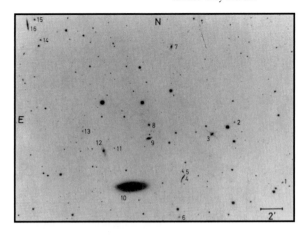

Fig. 3-24. The bright galaxy NGC4448 at a redshift cz = 693 km/sec surrounded by non equilibrium companions having redshifts from 5,200 to 36,000 km/sec.

The Referees Go Ballistic

When this new paper, with six authors, was sent to the Journal it elicited furious rejections by two referees in a row. Anonymous messages such as "ludicrous" and "bizarre conclusions based on an extreme bias of the authors wishing to find non-cosmological redshifts" were forwarded. One referee suggested that since we knew from the redshifts that the galaxies could not be interacting, the system should be adopted as a control for testing interaction evidence in other groups.

The principal author was so appalled he considered giving up research. But by a great stroke of fortune he asked for a third referee, who turned out to be a breath of sanity. Carefully enumerating all the ways in which this new study presented better observations than the previous ones, the last referee showed how the conclusions were properly drawn from the new data and also commented that the second referee seemed too angry to give a fair assessment of the worth of the paper.

Jubilation that the paper was finally published has to be tempered with the cold experience that much fewer than 1/3 of the referees in this field are objective. Disappointing also is the fact that even though this observational paper was published in a major journal, no notice was taken of it. I relate this story in detail because I think it reveals in the most telling way what the situation is in this particular branch of science. The facts can be consulted in *Astrophysical Journal.* 432,135 and references therein.

The Environs of the Average Bright Spiral

When pressed, skeptics usually complain that the examples are selected, that they don't represent a complete sample. But when a complete sample is carried out, for example a survey of 99 bright spiral galaxies carefully compared to non-galaxy control fields, and it shows interacting and peculiar companions are significantly associated with the central bright spirals (*Astrophysical Journal.* 220,47)—then the results are ignored. Not enough observing time was allocated to complete the measurements of the redshifts,

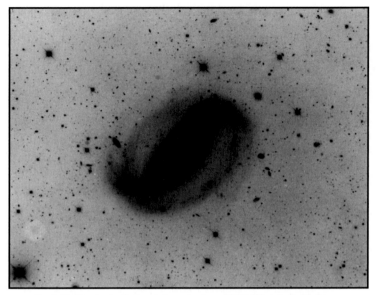

Fig. 3-25. The dusty, starburst, barred galaxy, NGC1808, is imbedded in a dense cloud of fainter galaxies which are undoubtedly of much higher redshift.

but that was really not necessary, as anyone could tell by looking at the galaxies that they were medium-high redshifts.

An example of such a galaxy is shown in Figure 3-24. The galaxy is NGC4448, and the analysis in *Astrophysical Journal*, 273,167 shows that the numerous, peculiar faint galaxies have redshifts ranging between 5,200 and 36,000 km/sec, while the central galaxy is at 693 km/sec. Another galaxy just embedded in a dense cloud of fainter, certainly higher redshift galaxies is the starburst, dusty NGC1808 in the southern skies. That is shown here in Figure 3-25.

The Origin of Companion Galaxies

The ejection of quasars from active galaxies documented in Chapters 1 and 2 leads to an extraordinarily important synthesis which I did not at first fully appreciate. It was not until after the later chapters on evolution of clusters of galaxies from clusters of quasars that I realized what the data did was to establish the origin of companion galaxies as the end point of the evolution of quasars!

To understand how we come to this result, one must go back to 1957 when Viktor Ambarzumian, from just looking at galaxies on Sky Survey photographs, proposed that young galaxies were born from material ejected from older, active galaxies. Independently I reached the same conclusion from my *Atlas of Peculiar Galaxies* in 1966. By 1969, the much respected Swedish astronomer, Erik Holmberg, was visiting the Mt. Wilson and Palomar Observatories in Pasadena. After 20 years of studying groups of galaxies, he was in possession of some startling evidence—namely, that companion

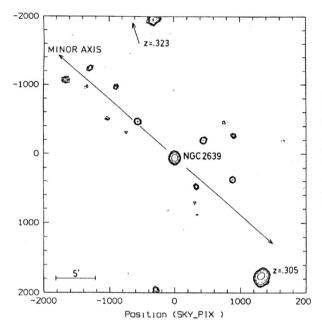

Fig. 3-26. X-ray map of the Seyfert galaxy NGC2639. A line of X-ray sources coming out exactly along the minor axis to the NE has four identified BSO's and Bcg's (quasars to be confirmed—see text for positions). The major pair of quasars measured by Margaret Burbidge is indicated at z = .323 and .305.

galaxies were preferentially distributed along the *minor* axis (rotation axis) of the dominant galaxy. As a young researcher at the Observatories, I discussed with him Ambarzumian's evidence for linear ejection of new galaxies, my evidence for ejection of radio quasars and pairs of objects across disturbed galaxies, and my most recent evidence (1969) that proto companions ejected in the plane of a spiral were stopped very close to the ejecting parent.

He agreed that his alignment of companions along the minor axis was strong evidence for the ejection origin of companion galaxies. But he would not utter a word of this at the Observatories for fear of being ridiculed. I was disappointed, because I badly needed support for my findings. Some time after he had gone back to Sweden, however, his paper appeared in his country's *Arkiv. f. Astronomie* To my delight, he forthrightly stated: " ... physical satellites of spiral galaxies are apparently concentrated in high local latitudes and ... favor systems which have [blue nuclear colors] and contain large amounts of gas. *The results seemingly point to one interpretation: that the satellites have been formed from gas ejected from the central galaxies.*" (Italics added for emphasis.)

What the X-ray quasar data showed in 1996, and what I did not immediately grasp, was that *the quasars were also preferentially ejected out along the minor axis!* This was first apparent in NGC4258 where the quasars were only 13 and 17 degrees away from the minor axis (Fig1-1). Then came NGC4235 (Figure 2-5) where the pair were only 2 and 12 degrees away from the minor axis of a clearly defined, nearly edge-on spiral. Finally NGC2639, pictured here in Figure 3-26, shows a group of seven X-ray sources coming out exactly along the NE minor axis. These latter, closer sources are apparently most recently ejected. The outer pair of quasars may represent earlier ejection when the minor axis was rotated in a somewhat different position. In general, such minor axis rotation could account for the greater spread in minor axis alignment of the older companion galaxies, as summarized in Table 3-1.

The fainter X-ray sources coming out along the NE minor axis of NGC2639 contain four optically identified BSO's or Bcg's. These blue stellar objects and compact galaxies are predicted to be less luminous, higher-redshift objects on their way out of the nucleus of NGC2639. In the naïve hope that they might someday be spectroscopically observed, I give their exact positions in Table 3-2.

It is not always possible to obtain cases of ejecting galaxies where the major axis (and hence the minor) is well defined. An example is NGC1097, a barred spiral, where the position of the major axis is necessarily uncertain by about 10 degrees (see Figures 2-7 through 2-9). Given this uncertainty, however, the four nearest quasars are within about ± 20 degrees of the estimated minor axis, as noted in Table 3-1. This places them just between the long, luminous optical jets which emerge from the nucleus and must

Table 3-1 Companion Objects around Spiral Galaxies

No.	Companions	$\Delta\Theta_1$	$\Delta\Theta_2$	$r_1 \sim r_2$	Reference
2	quasars across NGC4258	13°	17°	25-30 kpc	Pietsch et al. 1994
2 + (4)	quasars across NGC2639	0°	13°(31°)	10-400	Figure 3-26
2	quasars across NGC4235	2°	12°	500-600	Figure 2-5
4	quasars nearest NGC1097		~ 20°	100-500	Arp 1987
6	quasars nearest NGC3516		±20°	100-400	Chu et al. 1997
218	comp^{n}s around 174 spirals		~35°	40 kpc	Holmberg 1969
96	distbd. comp^{n}s around 99 spirals		~60°	150	Sulentic et al. 1978
115	comp^{n}s around 69 spirals		~35°	500	Zaritsky et al. 1997
12	comp^{n}s of M31		~0°	(700)	Arp 1987

Table 3-2 Properties of X-ray Sources in the NGC2639 fields

Name	X-ray ($ctsks^{-1}$)	R.A.	Dec.	Off axis	Ident.
Bright X-ray sources in Figure 3-26				(arcmin)	
RX J08443+5031	37.8	$08^h44^m19^s0$	+50°31'36"	20.6	QSO z=.323
NGC2639	13.5	8 43 37.9	50 12 19	0	Seyfert z=.011
NGC2639 U10	25.7	8 42 30.0	49 57 51	17.7	QSO z=.305
X-ray sources NE of NGC2639					
	2.4	8 44 46.1	50 22 54	14.9	BSO 19.2 mag.
	4.1	8 45 04.4	50 21 30	16.4	no ident.
	2.0	8 44 25.3	50 20 37	11.0	ambiguous
	1.3	8 44 48.7	50 20 34	13.8	BSO 19.9 mag.
	1.4	8 44 31.8	50 16 50	9.5	BSO 18.3 mag.
	2.6	8 44 07.2	50 16 28	6.0	BSO 18.8 mag.
	1.2	8 44 17.0	50 15 09	6.7	——

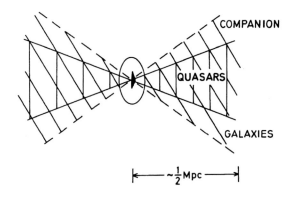

COMPANION

QUASARS

GALAXIES

$\longmapsto \sim \frac{1}{2}\text{Mpc} \longrightarrow$

Fig. 3-27. Distribution of companion galaxies and quasars along the minor axes of ejecting disk galaxies. The companion galaxies are at angles of approximately ±35 degrees (from Holmberg 1969 and Zaritsky et al. 1997). The quasar distributions are ±20 degrees from recent data discussed in this book. The observed size of galaxy groups is about 1 Megaparsec (3.26 millon light years).

represent some form of ejection.

The most compelling evidence for the origin of companion galaxies is certainly their coincident alignment with quasars. *Figure 3-27 shows how the quasars and companion galaxies occupy the same volume of space along the minor axis of the ejecting galaxy.* Together with the smaller, but systematic, excess redshifts of the companions, there seems to be no alternative to the conclusion that the quasars are ejected as more recently created matter, and that their intrinsic redshifts decay with time. The morphological and spectroscopic evidence shows them to be evolving into more normal galaxies. (Ejection along the minor axis involves no rotational component of motion and hence the objects remain on radial orbits as they age.) It will be discussed in Chapter 8 how the intervals of quantization of the quasar redshift values also decay into the smaller quantization values observed in companion galaxies.

We will discuss in Chapter 9 how the Narlikar/Arp application of the mass creation theory predicts initially rapid ejection of low-luminosity, high intrinsic-redshift objects, followed by a slowing and final stop out at about 400 kpc—just the range within which quasars and companion galaxies are found. As they continue to increase in luminosity, they slowly start to fall back, roughly (if not perturbed) along the line of original ejection. They also continue to diminish in intrinsic redshift as they evolve into normal galaxies, as shown in Figure 9-3. All these properties are observed—and cannot be explained on the assumptions of the Big Bang theory.

Spectacular Confirmation

As this book was being finished, word was received from Prof. Yaoquan Chu that he had measured with the Beijing telescope the new X-ray candidates around the extremely active Seyfert NGC3516. Fig. 11 of *A&A* 319,36,1997 shows the X-ray map derived by Arp and Radecke from the archive observations. There are five X-ray sources marked there, which Chu confirmed as quasars. Figure 9-7 shows their redshifts. A quick check of the minor axis direction revealed they all lay within about ±20 degrees of the minor axis. Together with the bright BL Lac type object to the NW, that meant *six* quasars coming out along the minor axis of this violently active Seyfert, all more closely aligned than the average of the many Holmberg companions.

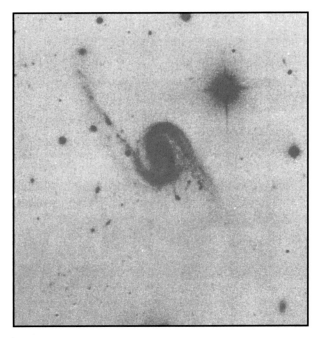

Fig. 3-28. *Atlas of Peculiar Galaxies* #65 showing companion galaxies on the ends of two long straight arms strongly suggesting an ejection origin.

But this did not exhaust the dynamite in Chu's observation. It turned out that the quasars were ordered, with the highest redshifts closest to NGC3516 and the smallest redshifts furthest away. Moreover, the redshifts were all very close to the quantized redshifts to be discussed in Chapter 8. I think the reader can already sense the exultation with which we received this news. Here was an observation which fulfilled every prediction as discussed in Chapter 9, was an incontrovertible confirmation of the sum of past observations, and which we knew eventually would ensure that 30 years of struggle would be of value.

Companions Ejected in the Plane

In the beginning of this chapter, I mentioned that it was my initial idea from studying photographs that if protogalaxies were ejected in the plane of their originators, they would pass through a phase of being companions on the ends of spiral arms. This idea was abetted by my belief that spiral arms were the result of ejection processes and that companions on their ends were related to the large "knots" one often saw along spiral arms.

Figure 3-1 shows a compact, luminous object emerging from the center of a galaxy trailing material behind it. Figure 3-28 shows two small companion galaxies on the ends of two long straight arms. Both of these pictures are from the Arp *Atlas of Peculiar Galaxies*. This means that already in 1966, we had pictures which showed at a glance that galaxies ejected compact objects which evolved into companion galaxies. Because knots in spiral arms were usually dominated by glowing HII regions, they were presumed to be excited by hot, recently formed stars. Was there something faint, of higher redshift

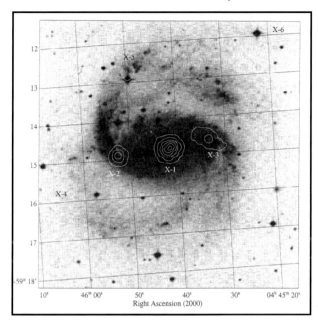

Fig. 3-29. The barred spiral NGC1672 has strong X-rays coming from its Seyfert nucleus. The diametric pair of X-ray sources across its nucleus suggests a pair of objects has been ejected in the plane of the galaxy and slowed down by interaction. (Picture from W.N. Brandt, J.P. Halpern and K. Iawasawa)

inside, that was masked by this gas of the parent galaxy? Or had the bullet passed on out, leaving star condensation to take place in the entrained gas of the galaxy (perhaps constrained in the magnetic tube of the spiral arm)? To answer such questions required observational hard work, which was obviously not forthcoming.

But the broad thrust of the observational inferences was helped by the X-ray observations reported in the first two chapters. There we saw that the newly created quasars which passed far outside the bounds of the galaxy had a strong tendency to lie along the rotation axis—or at least not in the plane. Were there any examples where the X-ray ejection had gone off in the plane? There may be some which have not yet been recognized, but one clearly probable case was called to my attention by the Japanese astronomer Awaki.

Figure 3-29 shows the barred spiral NGC1672. This galaxy has strong X-rays coming from its Seyfert nucleus, as well as X-ray sources coming from two diametrically opposite points, just at the ends of its bar where the curved spiral arms begin. We know that X-ray sources are ejected from the nuclei of active galaxies. What happens when they are ejected in the plane of the galaxy? Whatever their nature, they will be slowed down more going through the material in the plane than if they were ejected out of the plane. That means they will go through their rather rapid initial evolution closer to their galaxy of origin. If they evolve completely into a companion galaxy, they can then become higher-redshift companions connected to, or still interacting with, their galaxy of origin.

What do we see at the position of the two X-ray sources in NGC1672? Not much on routine low-resolution images—just the high surface brightness of the bar. But galaxies typically contain a lot of obscuring dust in the plane, particularly barred spirals

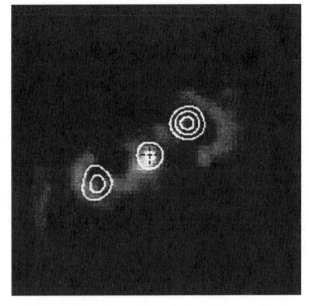

Fig. 3-30. Space telescope picture of the Seyfert 2 galaxy Mark573 by Wilson, Falcke and Simpson. Contour lines represent positions of radio sources. Notice Hydrogen alpha gas of the galaxy is drawn out along line of radio ejection.

that often have thick lanes of dust running out along the bar (See picture of NGC1097 in Figure 2-9). The Japanese satellite telescope, ASCA, which detects higher energy X-rays, registers the western source much stronger than the lower energy X-rays of ROSAT. This implies very strong dust absorption. If there were a highly obscured BL Lac object at the position of X-3 in Figure 3-29, how would we detect it? Even with advanced infrared equipment on large-aperture telescopes, we could have trouble identifying a faint object and getting a definitive spectrum. But that is not to say we should not try—eventually we should succeed in identifying what those strong X-ray sources are.

Another example of what I would take to be ejection in the plane is shown in Figure 3-30. The Space Telescope photograph of the Seyfert 2 galaxy, Mark573, shows a pair of radio sources ejected in opposite directions from a radio nucleus. Hydrogen alpha gas seems to form bow shocks around these ejected sources. But material from the galaxy is clearly drawn out in these ejections.

It Almost Never Happened

As important as I believe the intrinsically redshifted companion galaxies are to understanding the nature of cosmic redshifts, I must recall that I almost did not have the chance to publish or follow up the implications. It was 1967, and I had just finished *The Atlas of Peculiar Galaxies*. I had used my staff member's observing time to study the best examples of companions on the ends of spiral arms, and I submitted a paper, the abstract of which is reproduced above:

It was well understood at that time that the journal in which important papers were published was the *Astrophysical Journal*. The long-time editor of that journal was Subrahmanyan Chandrasekhar, a theoretician of great renown and generally considered

Companion Galaxies on the Ends of Spiral Arms

H. ARP

Mount Wilson and Palomar Observatories, Carnegie Institution of Washington
California Institute of Technology

Received June 24, revised August 11, 1969

Photographic and spectroscopic observations are presented which show that companion galaxies on the ends of spiral arms of normal galaxies tend to have (1) high-surface brightness, (2) emission lines characteristic of excited gaseous material, and (3) early-type stellar absorption lines in their nuclei. One companion is shown to be expanding. Another is shown to be probably receding from the center of the larger galaxy.

The hypothesis advanced is that these companions have been recently ejected (10^7—10^8 years ago) from the parent galaxy. It is concluded that they are short-lived, and that many are now in the process of expanding and ejecting secondary material. Holmberg previously concluded that small companions were found only along the minor axis of spiral galaxies because their disks stopped ejection in the plane. The companions on the ends of spiral arms in the present paper are considered to be examples of this stopping mechanism. It is further suggested that ejection of material through the disks of rotating galaxies is generally important in the formation of spiral arms.

Key words: galaxies — spectra of galaxies — companion galaxies — peculiar galaxies — ejection

a tough but fair guardian of its reputation. I don't know now how I ever could have imagined that he would have been pleased by these interesting new observational results. He returned the paper with a handwritten message scrawled across the top: "This exceeds my experience."

It took a little while for it to penetrate my stunned senses that he had rejected the paper without ever sending it to a referee. I suddenly felt a cold shudder of apprehension as I realized that my prospects in astronomy were not very bright if I had alienated the editor of the *Astrophysical Journal.* What to do? First I felt I had to safeguard the observations by getting them published somewhere they could be read and referenced. The only possibility seemed to be the journal that was just starting up in Europe called *Astronomy and Astrophysics.* With some trepidation, I submitted the paper there. After some anxious weeks the paper came back. A new jolt of panic hit when I saw it had been refereed by another renowned and conservative astronomer, Jan Oort.

Forcing myself to read on I was overjoyed to find that, although he did not agree with the interpretation, he found the observations valuable and interesting and accepted the paper for publication. In the ensuing years I came to know Oort better and found him to be an extraordinarily polite and gracious man. Underneath, however, he had opinions of steel, and apparently would never for a moment entertain a solution which violated the usual assumptions of astronomy. Many years later when he was nearing 90, after a warm dinner at his house, he wrote me a letter urging me to give up my radical ideas and once again participate in the privilege of doing mainstream astronomy. I thanked him and answered him with a quote from my wife: "If you are wrong it doesn't make any difference, if you are right it is enormously important."

A most vivid memory I have, however, comes from the time I was sitting next to Oort in the Krakow meeting of the International Astronomical Union. Ambarzumian was chairing the session and Oort leaned over to me and whispered: "You know, Ambarzumian was right about absolutely everything!" Many times since then I have wondered whether, if Oort had said that out loud, and backed it with his enormous influence, the paradigm of astronomy today might not be much different. And I wondered too whether this was not his real, intuitive intelligence slipping for an instant out from behind the secure conformity of accepted dogma. At any rate, although it pained him very much to see an interpretation given which was contrary to his own, it never occurred to him to prevent another genuine observer from speaking or publishing that opinion. To say that this was the ethics of an old-fashioned gentleman is to emphasize that ethics have changed today.

But that did not solve my problem with the *Astrophysical Journal.* The Director of the Mt. Wilson and Palomar Observatories called me down to his office. To my horror there was a copy of my paper sitting on his desk. Chandrasekhar had sent him a copy of my paper complaining that I had been caught up in a "phantasmagoria" (who could forget that word) and suggested that my Director do something about it. He did. He told me that my appointment would not be renewed next year.

Stunned disbelief and fright was my reaction. My understanding was that, though unwritten, my tenure at the Observatories was permanent. And yet what could I do if it were not? I could only mutter weakly that I would wait for his notification in writing. As the weeks and then months went by with no letter arriving, my terror began to subside and I began to think the crisis had passed. But I felt hunted, and there loomed the question of how I would handle the publishing of future observations.

In the height of the storm, there seemed only one principle to cling to—that was fairness. I knew the observations were good and the interpretation was based on scientific reasoning. The *Astrophysical Journal* had a responsibility to communicate them to other astronomers. Even though it would exacerbate my position, I decided I must protest to the Editorial Board. Almost a year passed and one day I heard that Chandrasekhar, after long and honorable service, had decided finally to relinquish the onerous burden of the editorship of the *Astrophysical Journal.* By then I was concentrating on further observational programs and I remember thinking: "...well, it is a hard job and he has been at it a long time, I suppose this had to come sooner or later."

A few months after that I came down to the Friday afternoon astronomy luncheon at Cal Tech. There was a seat open next to Fred Hoyle at the middle of the long table. I sat down next to him and started chatting happily about new observations. After a while Chandrasekhar, there on a brief unannounced visit, slowly entered the room and proceeded to the only empty seat at the table, directly opposite me. After finishing the subject with Fred I found myself looking directly across at a silent Chandrasekhar. Merely to make polite conversation I remarked: "You must be enjoying the respite from your arduous duties as Editor."

Suddenly there was one of those complete silences, as all conversation stopped and the whole long table turned to stare directly at us. Chandra rose up a few inches from his chair and said angrily:

"How could I continue to be Editor when people like you complained about me?"

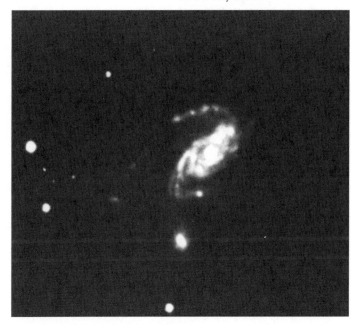

Fig. 3-31. A galaxy from the ESO Catalogue of Southern Galaxies, ESO 161-IG24. Companion galaxies appear to be attached to the ends of three spiral arms! Detailed spectroscopic observations would be extremely interesting.

I was stricken with embarrassment, but for the first time before or since, managed to come up with an immediate reply:

"I would hope, in spite of our professional differences, to remain cordial in public."

The table went back to conversation and we did not speak to each other for the rest of the meal. Come to think of it, we have never had an occasion to speak since then. In fact, these were the only words we spoke to each other in our entire lives.

Of course Chandrasekhar went on to be awarded the Nobel Prize for his work on structure of stellar interiors and allied subjects. About that time I was at an astronomical meeting and attended a lecture he gave. I was amazed that he spent a great part of the lecture talking about his relationship with his erstwhile teacher, Sir Arthur Eddington. In those remarks, Chandrasekhar stressed the emotional hurt that he had received when Eddington had strongly rejected his ideas on the degenerate cores of white dwarf stars. He emphasized what a debilitating effect it had on his outlook for a long time afterward. I was surprised, but I admired him for being able to talk about it publicly. Although at the same time it was sad to realize that he had then turned around and passed on a similar blow to someone else.

A Chance Galaxy

One day I was passing the photographic laboratories at the European Southern Observatory and I saw a pile of photographs they were discarding. I picked out the object shown in Figure 3-31. It turned out to be an object in the ESO Catalogue of Southern Galaxies, ESO 161-IG24. Just a chance galaxy. But it is so eloquent. Three spiral arms with a companion on the end of each arm. And what is more, the longest arm has a series of large knots along it, which look simply like nascent companions. Of course, it would be fun to examine this system with high resolution and spectra. But is it really necessary in a broader sense? The two major companions are obviously not falling in, and from what we know about ejection in many other galaxies, it is just inviting us to fill in the evolutionary links. Someone who knows galaxies will someday identify and observe it. Meanwhile, we can move on to investigate the questions of fundamental physical processes.

Chapter 4

INTRINSIC REDSHIFTS IN STARS!

If we are to believe the previous three chapters, then most extragalactic objects have intrinsic redshifts—ranging from large values for high redshift quasars and continuing right down to small values for low redshift galaxies. It was important to discover that low and medium redshift galaxies also had non-velocity redshifts because it meant that the effect pertained not only to quasars, which could be argued to be exotic and not well understood. Now the phenomenon could also be studied in nearby galaxies having gas, dust, and stars which could be resolved individually—all components about which we thought we knew most of what was important.

The Magellanic Clouds

The two nearest galaxies to us were reported as faintly luminous clouds in the Southern Hemisphere by early explorers. Even with the 74-inch telescope in South Africa in 1955, I was able to measure 10 magnitudes fainter than the brightest stars in the Small Magellanic Cloud (SMC). But it was not until 1980 that the brightest supergiant stars in both Clouds were measured with high spectroscopic dispersion.

Now the Magellanic Clouds are members of the Local Group of galaxies and therefore have an intrinsic redshift relative to the oldest galaxy in the Group, M31. (They also have positive redshifts with respect to our own Milky Way Galaxy, which would mark them in turn as our younger offspring). One could not help wondering whether the gas, dust and stars in these smaller neighbors all shared this same excess redshift—particularly the supergiants which are short-lived and must be, in some sense, younger than the rest of the galaxy. I remember vividly when, long ago, I first checked the companion galaxies to see if they were redshifted with respect to the dominant galaxy. It was with the same sense of not-daring-to-hope that I now approached the necessity of checking the supergiants in the Magellanic Clouds to see if they were, by any chance, redshifted with respect to their own galaxy.

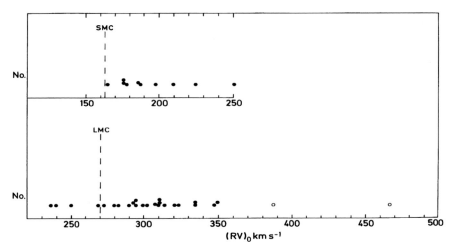

Fig. 4-1. Redshifts of supergiant stars in the Small Magellanic Cloud (SMC) and
the Large Magellanic Cloud (LMC). Corrected for stellar wind outflows, it is seen
that these most luminous stars in our nearest neighbor galaxies are
systematically shifted with respect to the mean.

They were! The accurate redshifts tabulated by John Hutchings in his study of
their mass-loss properties revealed at a glance that there was a systematic redshift. One
just had to run one's eye down the column of redshifts! Of course the mean redshift of
the gas in the Magellanic Clouds was known to within 2 km/sec. The mean redshift of
the older stars, though less accurate, agreed very well with the gas. Only the bright
supergiants were systematically redshifted, and it would be absurd to suppose that just
these stars were running away from their galaxy in the direction we happened to be
looking.

When I showed this result to some knowledgeable colleagues, they immediately
responded that you could not trust redshifts of stars with mass-loss winds. They were
sure that what I was seeing was an effect of the velocity of those winds on the measured
redshift.

> **Stellar Winds**: Bright supergiant stars pour out so much radiation that the
> pressure causes "mass loss" winds to leave the surface. As we look at the star, the
> intervening cooler gas which causes the absorption lines in the spectrum is moving
> toward us and therefore has a negative spectral shift.

I did not know much about stellar winds, so I had to draw a simple picture many
times over to convince myself that the redshifts measured from absorption lines coming
toward us would have to be made even *more positive* by correcting for this negative
velocity component. Since the spectral characteristics had been correlated with the
speed of the stellar wind in these supergiant stars in the SMC and LMC, it became
possible to correct numerically for the outflow velocity. When this correction was made,

all 10 supergiants in the SMC and 20 out of 24 in the LMC were positively redshifted with respect to their home galaxies. This is shown in Figure 4-1.

This is a devastating result, because one need only look at the distribution of redshifts to realize this could not be a chance occurrence. My colleagues still insisted that I must be applying the wind correction in the wrong sense. So I consulted the world expert in stellar mass loss, Rolf Kudritzki, who conveniently happened to direct the Munich Observatory. He was very helpful, invited me over to give a colloquium to his group, and verified that the corrections were being correctly applied. In fact he dug into his group's observations to supply me redshifts on additional supergiants that were made at the base of the photosphere where the wind was just beginning to accelerate to its final velocity. These turned out to be independent checks on the positive stellar shifts.

At this point I recalled sitting in Bart Bok's astronomy course at Harvard and listening to him describe the mysterious "K" effect. This is the effect that W.W. Campbell discovered in 1911, namely that the bright, blue stars in our own Milky Way galaxy had systematically higher redshifts than the rest of the stars. With my undergraduate self-confidence I set out to write him an equation that demonstrated how the young stars were streaming out from our galaxy. With some difficulty and a great deal of patience, he finally convinced me they should not be streaming away from just our sun's position on the edge of the galaxy. So gradually the problem receded from the front of my mind. But empirically this is the same effect that I had just found, forty years later, in the Magellanic Clouds.

Actually what had triggered my memory was a paper by a Canadian physicist, Paul Marmet. He was arguing the case that photons travelling through interstellar and intergalactic space would lose some of their energy by inelastic collisions. He had made quite a nice summary of the old evidence for the K effect in arguing for the tired light origin for redshifting of galaxies and stars.

Tired Light

Over the years, many people have argued that photons lose energy on their long voyage through space. This is an entirely reasonable idea, since the distances are the largest we have experience with. But there are several observational arguments that persuade me that this is not an important part of cosmic redshifts:

The first is that as we look to lower galactic latitudes in our own galaxy, we see objects through an increasing density of gas and dust until they are almost totally obscured. No increase of redshift has ever been demonstrated for objects seen through this increased amount of material. Secondly, we have seen that if we look through extragalactic space, the example of quasars linked to low-redshift galaxies demonstrates that two objects at the same distance with closely the same path length can have vastly different redshifts.

Finally, if we say there are clouds of a redshifting medium around each individual object, then there should be gradients of redshift across resolved objects, which are not observed. Further, we should see silhouetting and discontinuity effects between adjacent objects, which also are not observed. Perhaps on some level, light can get tired, but it does not appear to be significant in the redshifts we are dealing with.

The K Effect

When I realized that the excess redshifts of young stars in the Magellanic Clouds furnished a confirmation of the K effect in the Milky Way, it reminded me that no satisfactory explanation had ever been advanced for the phenomenon. When I went back over the literature, it was clear that the lack of explanation had gradually led to a disregard of observations having to do with the problem.

Plaskett and Pearce in 1930 and 1934 had tried to explain it as streaming motion, but it had to be over a huge sector of more than 120 degrees in the sky. Smart and Green in 1936 concluded "...K must be regarded as a systematic correction to the radial velocity of B-type stars...." Robert Trumpler, a well-known galactic astronomer at Lick Observatory, took a different approach, referring the redshifts of OB stars, not to a large volume of other stars, but to the redshifts of the young clusters to which they individually belonged. In 1935 he reported an excess redshift of 10 km/sec from a sample of nine of the most luminous stars in six clusters.

Now Trumpler believed he had an exciting confirmation of the much-publicized theory of general relativity (GR). But when the strength of gravity at the surface of these stars was calculated, it was found to be much too weak to give a gravitational redshift as large as observed. The rest of the astronomical community promptly forgot the K effect. But Trumpler went on believing it to be a gravitational effect, and continued to quietly measure more stars. In 1955, he presented his accumulated results in a conference in Bern on the 50th anniversary of relativity theory. There were 18 stars in 10 clusters which gave a mean excess redshift of +10 km/sec ±1 km/sec. This result had a chance of about one in 300 billion of being accidental.*

But we should also note that Trumpler measured his O stars relative to the early B stars in the cluster. As Finlay-Freundlich pointed out, the B stars themselves have a K effect, which approximately doubles the net excess of the O stars reported by Trumpler. Many other investigators had found this same effect in other associations of young luminous stars. For example the renowned stellar observer Otto Struve had shown that these types of stars in the Orion Nebula had excess redshifts of +15 km/sec. When corrected for stellar winds, these numbers were close to the average values found for the SMC of +34 km/sec and +29 for the LMC.

---

* Most astronomers have never heard of this result. The only reason I know about it is that Jurgen Ehlers was at the meeting and gave me the reference: *Helvetia Phys. Acta Suppl.*,IV,106,1956. It is interesting to ponder the implications.

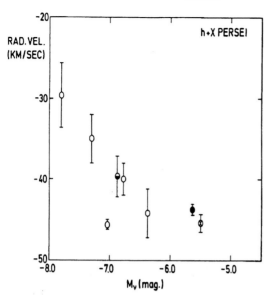

Fig. 4-2. Redshifts of the most luminous members of the h + chi Persei star clusters in our own galaxy. Open circles are luminosity class Ia, half filled circles class Iab, filled circle Ib and circled cross class MIa-b. Again the youngest stars are systematically redshifted. Wind corrections should accentuate the effect.

h + Chi Persei

Considering more recent results, the most prominent aggregate of supergiant stars which springs immediately to mind is the double cluster that can be seen with the naked eye as two faint smudges in the constellation of Perseus, *i.e.* h + Chi Persei. It is in the next spiral arm outward from our own, and contains some of the brightest blue and red supergiants known in our galaxy. After I consulted the latest measures on these stars by Roberta Humphreys, it was obvious to me that the brightest among these young stars were on average the most redshifted.

Figure 4-2 shows how the brightest have again about +15 km/sec excess redshift from stars only a little more than two magnitudes fainter. These fainter stars themselves undoubtedly have some K effect, and of course the mass loss correction which must be applied is even stronger for our galaxy because of its relatively high metallicity. Altogether, the magnitude of the excess redshift for the brightest stars in this best-known galactic cluster must be in excess of +30 km/sec, quite like that which pertains to the Magellanic Clouds.

As a side comment, we should note that since these corrections have not been taken into account, the spiral arm kinematics derived for our own galaxy must, to some extent, be wrong and should be recalculated. It is interesting that when I sent these results to the researcher responsible for measuring the h + Chi spectra and asked for comments, there was no response.

K Effect in Other Galaxies

The K effect can be tested in other nearby galaxies if there is spectroscopic evidence of young, high-luminosity stars. What shows up in the composite spectrum is absorption lines of hydrogen, particularly the higher excitation lines which are sharper in

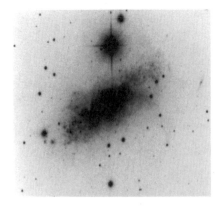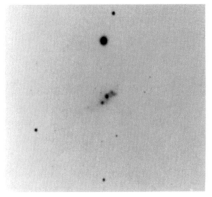

Fig. 4-3. The deeper exposure shows the dwarfish morphology of this nearby, active galaxy. The weaker exposure shows the two clusters of young stars which dominate the interior of the galaxy. These clusters are redshifted by 36 and 35 km/sec with respect to the mean of the galaxy.

the rarified atmospheres of young giant stars. As an example, a nearby dwarf galaxy called NGC1569 is shown in Figure 4-3. On one of my last runs at Palomar, I took spectra of the two stellar-appearing objects in the center of this peculiar dwarf. Gerard de Vaucouleurs had thought they were radiation from high-energy electron gas (synchroton sources), but to my amazement they turned out to have sharp line spectra exactly like a very bright supergiant.

In Figure 4-4, the spectrum of the brightest one of these objects is shown to have such a low-density atmosphere that the hydrogen lines are so narrow they can be easily seen down to H12. In lower luminosity stars one does not see the hydrogen series so narrow. Composite spectra of average galaxies principally show just the H and K lines of Calcium.

Under higher resolution these objects later turned out to be compact star clusters** in which the stellar orbital velocities were not high enough to broaden the spectral lines. (It is intriguing that these clusters are double, as is the h + Chi Persei cluster in our own galaxy. There are many examples of double cosmic objects, but the question of why has not even been asked, much less an answer attempted.)

With such good spectral resolution, the narrow lines can be measured accurately and compared to the accurate radio measures of the neutral hydrogen in which most young galaxies are imbedded. In the case of NGC1569, as Table 4-1 shows, the bright stars in this young double cluster are just about +35 km/sec higher redshift than the hydrogen in the galaxy. This is the same situation as in the Magellanic Clouds. And, of course, the mass loss corrections could make this excess redshift even larger.

Finding obvious evidence for intrinsic redshifts in a number of independent analyses of objects which we think we know as much about as stars is a sensational

---

**   When we wrote a paper about this observation Sandage voted for the star cluster interpretation and I voted for single stars: Sandage turned out to be right.

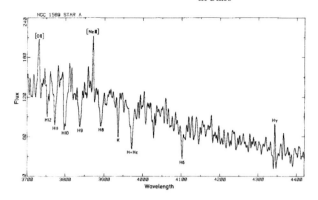

Fig. 4-4. The spectrum of the brighter star cluster in NGC1569, showing the narrow hydrogen absorption lines which mark the stars as evolutionarily young.

development. Moreover, as we shall see later, the relation between young age, low mass-to-luminosity ratio and increasing redshift is crucial to discovering the cause of the intrinsic redshift.

It is not a Discovery until it is Communicated

Gathering all the above material on stellar excess redshifts together has furnished quite an impressive demonstration of the effect. The excess redshifts of the Magellanic Cloud Supergiants were especially striking, since the mean redshifts of the galaxies were so accurately known. Consequently, I thought it appropriate to submit it to the French editor of *Astronomy and Astrophysics*, because he was a specialist in the Clouds and had just finished an extensive study of the hydrogen gas in them.

How naïve of me! The paper came back with not just a rude, but a savage rejection. To answer one objection, I undertook the enormous task of differencing the redshift of each supergiant with the local hydrogen in its immediate vicinity. Of course, the result was the same. The paper came back with an even stronger rejection and the

Table 4-1 Intrinsic Redshifts of Stars in Nearby Galaxies

Galaxy	Objects	K effect + Mass Loss	= Total (cΔz) (km/sec)
Milky Way	O-B stars	0.6 to 22 + (17)	=28
" "	H + chi Per	(15) + (17)	=32
LMC	Supergiants	7 + 22	=29 ± 6
SMC	Supergiants	17 + 17	=34 ± 8
NGC 1569	Cluster A	36 + —	≥36 ± 17
"	Cluster B	35 + —	≥35 ± 22
NGC 2777	early integ. Sp.	31 + —	≥31 ± 8
NGC 4399	" " "	25+ —	≥25 ± 15
M31	Irreg. blue vars.	(100) + —	≥(100)
M33	" " "	21 + —	≥21

See *Mon. Not. Roy. Astr. Soc.* 258, 800 and *Ap. J.* 375, 569 for analysis suggestion that I should have presented it as proof of the incorrectness of the spectroscopic redshift measures.

The aspect that upset me the most was that this very important scientific data was being censored by an editor whose primary responsibility was to communicate such data. After some anguished thought, however, it occurred to me that establishing the principle that it was the foremost obligation of the editor to publish valid scientific data was even more important than communicating the data.

As in other cases, I convinced myself that if people faced with clear cases of improper conduct did not take a stand, there would be no hope for reform, and in fact, matters would probably get worse. So I set about the daunting business of finding out who the editorial board was, writing a summary, including the pertinent materials and making an official protest. It turned out the then Director of the Institute where I had taken refuge as a guest scientist was at the meeting of the European Council which considered the complaint. He never said anything about it to me, but one of my other colleagues heard about it and was furious at me. As far as I know, my protest did not accomplish anything. But I did find out that the editor in question was on the visiting committee to my host institution. I was pretty miserable about the whole affair, but I still felt I had done what I had to do.

All that, of course, had not solved the problem of communicating the data. With trepidation I then submitted the paper to *Monthly Notices of the Royal Astronomical Society*. Great day! I got a referee who, while a little grumpy, accepted the paper. I felt it made an impressive case when finally published (MNRAS 258,800). But that, I thought, was that—another important development dropped into the black hole of theory.

I was astounded some time later when I got a telephone call from the foremost expert on luminous stars in the Milky Way, Adrian Blaauw. As an older astronomer, he harked back to the era when astronomers studied and really knew about stars in our own galaxy. He said: "That paper should have been written ten years ago." I felt euphoric. He came to my office later and we talked at length about it. He got some of his students to invite me to give a colloquium, but most of them sat warily silent, and one or two tried to rework the observations to fit the conventional assumptions.

After that it slid downhill in a way that is instructive to recount. Two of the modern astronomers still working on the K effect were measuring less luminous O and B stars (they had run out of the most luminous ones). Moreover, they did not correct for wind effects, so they naturally found a smaller effect. They then made the most satisfying of all announcements: "With the measurement of a larger sample the anomalous effect has gone away."

About that time Geoffrey Burbidge was starting to include the K effect results in his lectures as further evidence for intrinsic redshifts. In one lecture he gave, a young astronomer stood up in the audience and said: "Oh these positive wind effects have been observed in the sun and they are easily explained." I talked to him after the lecture and he did not have any understanding of either the sun or the supergiants. After another lecture at the IAU, one of the astronomers who had come to the wrong

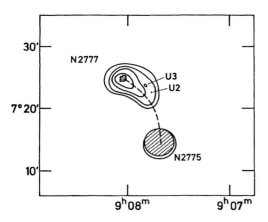

Fig. 4-5. NGC 2775 is a large Sa galaxy which, as the hydrogen gas contours show, appears to be ejecting the star burst companion NGC2777 which has +139 km/sec greater redshift. U3 and U2 are ultraviolet rich objects whose spectra have not yet been measured.

conclusion from a test of the Trumpler effect with fainter stars came up to Geoff and said: "Old friend, after making the most thorough modern investigation, I have proved that the K effect doesn't exist." Now Geoff has been a real hero in forcing unwilling audiences to listen to evidence for non-velocity redshifts, but it is understandable that he stopped including the K effect in his presentations.

The Pleiades Manuever

Often when a proponent of a cherished viewpoint is threatened by contrary evidence, he calls for further observations. This delays decision, but is an unassailably proper scientific position. It is apt to work out, however, in the following way: Let us say someone believes in a completely homogeneous universe, and the Pleiades star cluster is an embarrassing observation. He says there is too small a number of stars to be statistically significant—so let us test this hypothesis by getting a larger sample. He then measures all stars down to 21st magnitude (non-cluster, background) and reports triumphantly that the clustering has decreased to the point of insignificance!

That leaves the average non-specialist looking up at the sky and wondering how that clump of bright stars could have been accidentally so conspicuous.

NGC2777

Deserving of special mention is this system, which contains a main galaxy, NGC2775, and a companion galaxy, NGC2777. The companion has very accurate measurements from its early-type stellar spectrum. The system is sketched in Figure 4-5, and the spectrum of NGC2777 is shown in Figure 4-6. In this spectrum the metal-indicating K line is practically absent, *marking the galaxy as so young that successive generations of stellar evolution have not had time to enrich the metal content.* Again we find the younger stars with an accurately measured excess redshift (K effect) of +31 km/sec (Table 4-1). Again, this does not have mass-loss correction added. (Although in metal-poor stars the atmospheric opacity is less and the stellar winds do not blow so hard.)

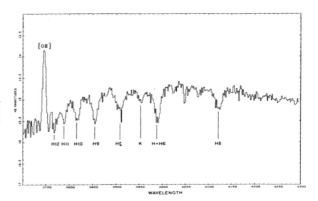

Fig. 4-6. The spectrum of NGC2777 shows the narrow hydrogen absorption lines of young supergiant stars. Exceptionally the K line of Calcium is very weak in this spectrum, indicating that the contribution of older stars in this starburst companion is negligible.

This system is a particularly powerful example of a companion with a higher redshift than its parent (+139 km/sec with respect to NGC2775, the large, neighboring Sa galaxy). The companion even has an umbilical cord, a streamer of neutral hydrogen (HI) leading back toward the larger galaxy. It is a quintessential example of a slightly younger, slightly higher-redshifted companion galaxy just now emerging from its parent galaxy (as described in the preceding chapter), but now we see a younger generation of stars within it having higher redshifts still.

Of course, the merger/collision school interprets this as a recent collision between two galaxies of the same age.

The Urge to Merge

As mentioned in a previous chapter, some astronomers saw peculiar galaxies neighboring each other, and immediately assumed that the peculiarity was caused by the galaxies falling into each other. By ignoring the empirical evidence for ejection from galaxies, they illustrated an unfortunate tendency in science, namely that when presented with two possibilities, scientists tend to choose the wrong one. We can explore the theory behind this observation later, but first let us examine the reasons why collision would be an unsatisfactory model for systems like NGC2775/2777.

1) The HI from NGC2777 leads directly back towards the center of NGC2775, implying the companion originated directly from that nucleus. Two galaxies falling together would have some transverse component of velocity and, therefore, not fall directly together but have a parabolic encounter.

2) Companions around a main galaxy would have to orbit for the order of 15 billion years and only occasionally fall in for an encounter. They would have to be like the huge Oort cloud of comets, which supply an occasional visitor to the inner planetary system. Large reservoirs of companions are not observed around central galaxies.

3) There is now simply an enormous amount of evidence that the companions are systematically redshifted. The merger/collision hypothesis would require as many companions approaching as receding.

Why not Publish it?

The paper titled "The Properties of NGC2777: Are Companion Galaxies Young?" took 2 years and 3 months to be published in The *Astrophysical Journal*. The referees started out with comments like: "crackpot theories", "twisted judgment", "quite nonsensical", "revolutionary conjectures" and "unsupported conjectures." After five revisions, formal complaints to the editorial board, and detailed editorial intervention to deal with complaints like "too hard to read" and "poorly organized," it was finally published. (My favorite complaint was that NGC2777 had not been identified as the companion—apparently this referee had not read the title or the abstract, or even looked at the figure.)

One referee contributed an engaging simile, however, when he objected that it was ridiculous to imagine new galaxies being ejected like "popcorn" from old galaxies.

The Problem of Star Formation

The field of research which the NGC2777 observations had so deeply offended was that of merger scenarios. But there was more to it than that, because NGC2777 was clearly a "starburst" galaxy in which rapid, current star formation was taking place. One of the major cornerstones of the conventional theory is that stars are formed as the result of collision of two gas/dust regions. (The idea is roughly that the gas would be compressed by the shock, and the globules would collapse to form stars.)

Now my co-author, Jack Sulentic, and I were disrespectful enough to suggest that smashing two lumps of gas together was the worst possible way to make stars. It just heated up the gas and generally caused condensations to dissipate.* We instead suggested that the key to star formation was to constrain the gas as it cooled, as in spiral arms of galaxies, which constrain ionized gas in the magnetic tubes which define the arms. Any directed ejections from an active nucleus would stretch out magnetic field lines from the interior into flux tubes. The rotation of the galaxy would turn these into spiral arms, and we were back to my old favourite: spiral arms as ejections which then became loci of new star formation. With amorphously shaped starburst galaxies, perhaps a rotational disk of material had not yet formed.

In attempting to publish the evidence for an ejection origin for the starburst companion NGC2777, we had trod on a number of sacred assumptions including mergers, merger-induced star formation, and all galaxies being old and having only velocity redshifts. It was flattering that the majority in the field tried everything to block a discussion of these problems, but it was also terribly discouraging that when it was finally published (*Astrophysical Journal*. 375, 569, 1991), absolutely no discussion ensued.

---

* This is reminiscent of the Chamberlain-Moulton Planetesimal Hypothesis, which was the accepted theory of formation of our planets in the early 1900's. The idea was that a passing star tore off a filament from our sun which condensed into the planets. An elementary school pupil could have said "But a hot ball of gas will dissipate not condense." I don't know if this was the cause of the theory's downfall, but it was subsequently replaced by planetary accretion of cold material.

Starburst Galaxies

Now that we have mentioned these galaxies which everyone agrees contain a lot of young stars, we should make some attempt to relate them empirically to other galaxies in the characteristic groups in which they, like most other galaxies, occur. This is done in Table 4-2, where all the best known starburst galaxies are listed. This way of doing business is not liked by academics. It is too simple—just list the most prominent examples and see if they give a clear trend. One may not recall every case, but if the trend is clear a few cases will not change it. If one goes to less prominent examples, they will probably also confirm the relation—and if they do not, one must check that they are the same kind of objects. This method just requires a good acquaintance with what the sky contains, and can be done initially in the head.

There is a clear trend. Table 4-2 lists the brightest and best-known galaxies which are classified as starburst (Am = amorphous morphology, usually high surface-brightness, young stellar spectral-type and "hot" infrared colors, which everyone agrees is a criterion of current star formation). The last two columns of the table then clearly show that these starburst galaxies are companions to some of the largest, best-known central galaxies in the sky, and they are almost always redshifted with respect to that galaxy.

Table 4-2 Nearby Starburst Galaxies

Galaxy and Type	Optical Spectrum	IR Flux Ratio (60/100 μm)	Companion Status	
			cΔz from Main Galaxy	ID
NGC 2777 Am	Very early absorption + emission	0.58	$+139 \pm 9$ ks s^{-1}	NGC 2775
M82 Am	Early absorption + emission	1.02	$+286 \pm 5$	M81
NGC 3077 Am	Early absorption + emission	0.59	$+57 \pm 6$	M81
NGC 404 S0$_{pec}$	Early absorption + emission	0.49	$+228 \pm 10$	M31
NGC 1569 S$_m$ IV (Am)	Very early absorption + emission	0.91	$+157 \pm 8$	M31
NGC 5195 SB0 (Am)	A-F	1.7	$+11 \pm 8$	M51
NGC 5253 Am	Early absorption + emission	1.06	(-104 ± 9)	NGC 5128
NGC 1510 Am	Early absorption + emission	0.63	$+69 \pm 9$	NGC 1512
NGC 520 Am	A-F	0.66	...	Uncertain
NGC 1808 SB$_{c\ pec}$	...	0.72	...	Uncertain
Dominant Galaxies in Groups for Comparison				
M81 S$_b$ I-II	Normal late-type	0.27	...	...
NGC 2683 S$_b$ I-II	Normal late-type	0.20	...	...
NGC 7331 S$_b$ I-II	Normal late-type	0.24	...	...

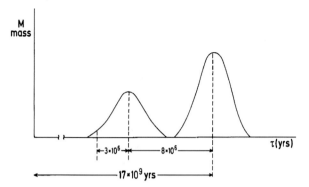

Fig. 4-7. Schematic representation of a massive galaxy created 17×10^9 years ago. The smaller companion galaxy was created 8×10^6 years later. The matter in the brightest OB stars in the companion is created 3×10^6 years after the epoch of creation of the companion.

Interim summary

The most important result so far is that all the empirical evidence discussed in this book establishes a pattern whereby a large, old galaxy has ejected younger material which has formed younger smaller companion galaxies around it. The younger galaxies in turn eject material, which forms even younger quasars and BL Lac objects.

The age hierarchy is evident from the properties of the objects in the characteristic groups in which galaxies occur. The ejection origin of the younger objects is evident from their pairing across active nuclei, their luminous connections back to active centers and the general propensity for ejection of radio and X-ray emitting material and excited gas.

In one-to-one correspondence to the age hierarchy, there is a redshift hierarchy. Every testable line of evidence shows that the younger the object is within the group, the higher its intrinsic redshift is. But now we have to deal more rigorously with what "young" means for stars and galaxies.

Ages of Stars

Conventional star formation takes place in a gaseous medium when a density fluctuation gravitationally contracts to the ignition point for nuclear fusion radiation. The gas could have been in existence for a long time, but the life of the star is viewed as starting at the time of ignition. If a star then evolves for, say, a few billion years to a fainter luminosity, it is usually thought of as an "old" star. We will retain the term "age" as an indicator of its evolutionary stage, but we also will introduce the concept of the age of the material out of which the star or galaxy is made. In the Big Bang theory, all matter was made at the same time 15 billion years ago. Thus, there is no concept of "young" and "old" matter in the current parlance. But here we will have to speak of a quasar being made out of young matter (more recently created matter) in order to account for its intrinsic redshift.

How would this work in the case of real galaxies, say M31 and our own galaxy, or the Milky Way and the Magellanic Clouds? Figure 4-7 shows a parent galaxy created 17 billion years ago. Unlike the current Big Bang theory, which assumes all matter was created instantaneously, we suggest it is vastly more realistic to create any given protogalaxy over a small interval of time—about 6 million years as indicated by the gaussian distribution in Figure 4-7. (The interval of creation would be .03% of the age of the parent.) Now suppose about 8 million years after the creation of the parent, the parent ejects new material which eventually evolves into a companion galaxy. As Figure 4-7 indicates, this younger companion has a similar spread in the age of its material. *Then the most recently formed stars in the companion will be preferentially composed of the most recently formed matter.*

This would account for the brightest stars in the Magellanic Clouds having small intrinsic redshifts and the brightest stars in our Milky Way galaxy having the same order of intrinsic redshifts. We will elaborate on this somewhat difficult point in a later section, but first we should explain that the empirical result that "younger" objects have higher intrinsic redshifts does have a theoretical explanation. The explanation, in fact, allows us to test the *numerical* relation between the age and the redshift.

Why does Younger Matter have Higher Redshift?

In 1966, when it became clear that the quasar 3C273 was in the Virgo Cluster and that other quasars were associated with nearby galaxies, the intimidating question of what caused their redshifts cast a foreboding shadow over everything. It quickly became clear that unless there was an acceptable explanation, the result would never be accepted no matter how strong the observational evidence was. And if the result was not accepted, the observations that would lead to an explanation could not be followed up. This is the insoluble dilemma which limits Academia.

But simple, empirical pattern-recognition in the observations did show that there was a continuous physical transition between the compact, high-redshift quasars through the high surface-brightness, active companion galaxies, and finally down to the more relaxed, normal-appearing galaxies. This, empirically, was also a continuous sequence in age from the youngest with the highest redshift, to the oldest with the lowest redshift. But what was the cause, what was the reason that the youngest had the highest redshifts?

In 1964, Fred Hoyle and Jayant Narlikar proposed a theory of gravitation (I would now prefer to call it a theory of mass) which had its origin in Mach's principle. According to this theory every particle in the universe derives its inertia from the rest of the particles in the universe. Imagine an electron just born into the universe before it has time to "see" any other particles in its vicinity. It has zero mass because there is nothing to operationally measure it against. As time goes on it receives signals from a volume of space that enlarges at the velocity of light, and contains larger and larger numbers of particles. Its mass grows in proportion to the number and strength of the signals it receives.

Now comes a key point: If the mass of an electron jumping from an excited atomic orbit to a lower level is smaller, then the energy of the photon of light emitted is smaller. If the photon is weaker it is redshifted. We will explore this rigorously in the Cosmology chapter, but it suffices here to understand that *lower-mass electrons will give higher redshifts and that younger electrons would be expected to have lower mass.* (Of course the masses of all particles scale together, but it is primarily a change in electron mass which determines the change in wavelength, or spectral shift, of the photon emitted in a transition.)

Quantitative Predictions

It was a great comfort to have the logical requirement that newly created matter be initially highly redshifted. This idea fitted the whole range of important data very well. Moreover, if the conventional Big Bang theory created matter once, there was no reason why matter should not be created again at later epochs. In fact, it would not strictly be new creation but merely materialization of mass-energy from a different, perhaps diffuse location.

However, the credo of science which is instilled in schools says: "Real scientific theories predict numerical results which can be measured—the more decimal points the better!" I was rather gloomy that the age-redshift relation would ever be formulated in enough detail in my lifetime to predict numerical relations.

Then something happened which at first I was not aware of. In 1977, Jayant Narlikar generalized the sacred equations of general relativity (*Annals of Physics*, 107, 325). The consequences, I believe, are profound, and we will discuss them at length in the cosmology chapter. But the essence of the solution was very simple:

(1) $m = at^2$ where a = constant

This means that the particle masses, m, vary as their age squared. Since the redshift varies inversely as the mass, we have a numerical relation between the age of a particle and its redshift:

(2) $(1 + z_1)/(1 + z_0) = t_0^2/t_1^2$

where z_0 is is the redshift of matter created t_0 years ago and z_1 is the redshift of matter created t_1 years ago. We take t_0 to be the age of the oldest created matter considered and, for reference, its redshift to be $z_0 = 0$. The first part of Table 4-3 lists the excess redshift

Table 4-3 Intrinsic Redshift-Age Calculations from Eq. (2)

cz_1 (intrinsic redshift) (km.sec)	$t_0 - t_1$ (younger age of matter) (yrs)	
12	1×10^6	Some evolved stars
35	3×10^6	Supergiant stars
71	6×10^6	Local Group companions
106	9×10^6	" " "
8,000	6.7×10^8	NGC 7603 companion
28,000	2.2×10^9	NGC 1232 companion

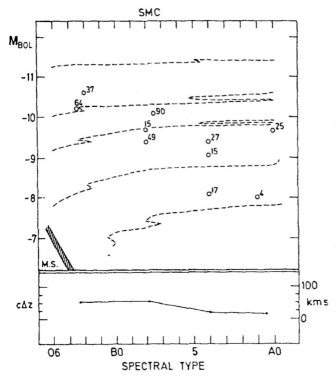

Fig. 4-8. The evolutionary tracks of high mass stars in the Herztsprung-Russell (absolute magnitude—temperature) diagram. The brightest stars evolve away from the main sequence (M.S.) in a time of only about 3×10^6 years. The open circles represent young supergiants in the the Small Magellanic Cloud. Their excess redshifts are labeled in km/sec.

calculated for matter which is from 1 to 9 million years younger than the matter comprising the parent galaxy.

It is apparent that for matter 3 million years younger than the average material in the galaxy, the intrinsic redshift is about 35 km/sec. Three million years is very close to what we would estimate for how much younger the matter is in the most luminous supergiants. How do we make this estimate? Looking at Figure 4-8 we see (essentially) the luminosity-temperature diagram for the SMC with the brightest supergiants plotted as open circles. The dashed lines indicate tracks along which they evolve rapidly. The stars traverse the uppermost track in just about three million years. Now stars made of matter of the average age of the galaxy will fill in a diagram with tracks of stars from a given luminosity faintward. Stars made of matter 3 million years younger will fill in these same tracks *plus a brighter track about 3 million years younger than the brightest old-matter track.* Since there is an upper limit on luminosity due to radiation pressure and speed of evolution, when we look at the most luminous stars in a galaxy we will see the stars made of the youngest matter. That age difference will be just about 3 million years. That predicts an intrinsic redshift of about 35 km/sec, which is just about the K effect we measure for the most luminous stars in the range of galaxies previously discussed.

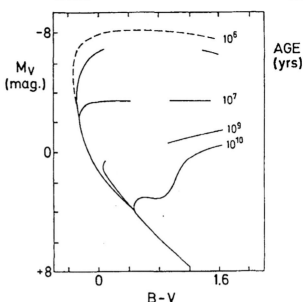

Fig.4-9. A schematic Hertzsprung-Russell diagram of a galaxy with sequences of stars of different ages labeled. Stars formed 3×10^6 years more recently than the mean age of the galaxy would only deviate from the brightest, youngest evolutionary tracks as shown by the dashed line.

The Ages of Companion Galaxies

What about the ages of the companion galaxies? We have discussed in previous chapters how galaxies, particularly in their young and active stages, eject objects and material which later evolve into companion galaxies. Later we will argue that it is actually the creation of new matter that causes the ejection. But regardless of the process of ejection, whether the ejecta proceed unimpeded along the poles and then fall back to a near orbit, or traverse the plane and emerge slowly, the material is at least somewhat younger than the parent galaxy.

How much younger can only be judged by looking at the stars in the companion. Figure 4-9 shows how the color-magnitude diagram of most of the stars in a galaxy would look if we could plot their individual luminosities and colors. The oldest stars have evolved to low luminosities and red colors. As we go to brighter absolute magnitudes, the stars become generally bluer until we encounter the brightest supergiants, which, as we saw in Figure 4-8, have very short lifetimes of only a few million years. But when we look at a galaxy that is too far away to resolve very faint stars, we have to deal with a composite spectrum that is an integration of all the stars into one spectrum.

If the composite spectrum is dominated by early-type stellar spectra, then we know the majority of the stars are young, and the galaxy itself is probably fairly young (barring a necessarily rare major burst of star formation in an old galaxy). If the composite spectrum is dominated by low luminosity stars, we know the galaxy is old, but it is difficult to say exactly how old. For example, if the last track, the 10 billion year-old evolutionary track in Figure 4-9, was missing, the composite spectrum would look roughly the same, and yet the galaxy would be much younger.

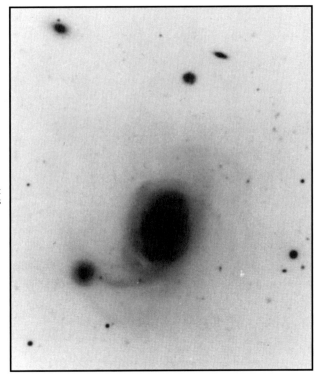

Fig. 4-10. The galaxy NGC7603 has a redshift of 8,700 km/sec. Linked to it is a companion with 17,000 km/sec redshift. It turns out that NGC7603 is a Seyfert galaxy.

If we look closely at the companion galaxies in our Local Group, the reddest, most dynamically relaxed dwarfs like M32 or NGC205 (see Figure 3-1) show some indication of more ultraviolet light, which indicates they are somewhat younger than an old giant E galaxy—but not much. If we say that they are between ten and a hundred million years younger than the parent M31, then they would be predicted by Table 4-1 to have between 100 and 1000 km/sec intrinsic redshift. This is about what is observed, as Figures 3-13 and 3-16 in the previous chapter demonstrate.

Of course, those with more indicators of young stars like NGC404, IC342 and M82 show higher excess redshifts than the older composite spectra. If we go to companions around higher redshift galaxies, like the companions around NGC7603 and NGC1232, we see from the bottom part of Table 4-3 that they would have to be about .7 and 2 billion years younger, respectively, than their parent galaxy.

The companion to NGC7603, as shown in Figure 4-10 (and Plate 4-10), is clearly linked back to the center of the active Seyfert galaxy; but it always bothered me that the integrated spectrum looked old. Now, however, we can see from Table 4-3 that it need only be .7 billion years younger—that is only about 4% younger than an assumed age of 15 billion years for its parent—which would be quite old enough to account for its spectrum.

Even in the NGC1232 blue companion (the companion is identified by a white arrow on the back cover of *Quasars, Redshifts and Controversies* and discussed therein on p.88), the required creation epoch is only about 2 billion years later (12%) than the creation of the parent spiral. In the NGC1232 blue companion, however, the spectrum is abnormally, peculiarly blue. Certain old star indicators like NaI and MgI absorption are missing—perhaps indicating that the era of the oldest star formation has been pushed to noticeably more recent times than in normal galaxies.

Both of the latter two objects are important objects to investigate with the high-resolution imaging and spectroscopy which is available from the world's giant new telescopes. They are both intrinsically fascinating, and their study would furnish the opportunity to quantitatively explore the time scales and physical processes of the creation of the most fundamental elements in our universe.

In total, there are already 38 known cases of high excess redshift companions as mentioned in Chapter 3. They range from +4,000 to +36,000 km/sec higher redshift than their parent galaxy. Anyone who looks at the examples in *Quasars, Redshifts and Controversies* or the original papers, *Astrophysical Journal.* 239,469,1980 and 256,54,1982—anyone who just looks carefully at the interactions and connections between these high and low-redshift galaxies knows they are physically associated. Most of the companions have spectra indicating unusually large components of young stars, and should have been investigated in the greatest possible detail long ago.

Does the Theory Explain the Observations?

In essence a theory merely connects together all the known facts in the simplest possible way. We have been driven by the observations to consider what would cause the redshift of galaxy material to decrease as it aged. The only simple possibility seemed to be that the masses of elementary particles increase with time. We have seen that this satisfies the fundamental constraints of physics as we presently understand the subject, *i.e.* it is a valid solution of the generalized Einstein field equations.

The observations which need to be explained are diverse, ranging from the low redshift, K effect excess of young stars, to the huge intrinsic redshifts of the quasars. The age of the matter making up the stars which show the K effect can be inferred from stellar evolution theory. The age-redshift formula then allows computation of their intrinsic redshift. It is impressive when the predicted redshift is observed to be in agreement with the observed redshift to better than an order of magnitude.

Aggregates of stars in the form of companion galaxies can have their ages estimated by the nature of their composite spectra. It turns out that these estimates predict their observed intrinsic redshifts to within an order of magnitude. In later chapters we will show that the very large redshifts of quasars are also predicted by the young ages inferred for them.

In summary, it is gratifying to note that we have used most of what is known about stars—their composition and structure, spectra, clustering, evolution and relation to galaxy morphology—to compute what is causing their non-velocity spectral shifts. This body of knowledge, taken as a whole, has pointed us toward a fundamental property of mass. It is difficult to imagine how such a wide range of observed phenomena could be explained so satisfactorily if this principle were incorrect. In the coming chapters it will be exciting to see what insights this gives us into the nature of creation and evolution in the universe.

Chapter 5

THE LOCAL SUPERCLUSTER

One of the most dedicated cataloguers of galaxies was Gerard de Vaucouleurs. In the 1950's, he began to notice that the bright galaxies fell along a great circle in the sky. He realized that the galaxies were distributed in a flattened disk, and that their strongest concentration was in the direction of the constellation Virgo. We were somewhere near the edge of this supercluster of galaxies and we saw its center as the Virgo Cluster about 17 Mpc (55 million light years) distant. Because it violated their assumption of homogeneity for the universe, other cataloguers initially scoffed. But now determinations of supergalactic latitude and longitude are an accepted way of locating objects in the Local Supercluster. It turns out, therefore, that the Virgo Cluster is the center of the largest physical aggregate of galaxies we can study in detail. We will see that it contains the greatest range of types of objects, ages and energies. It is there that we can best observe a wide range of physical processes and the relation of different objects to each other.

In Chapter 3 we saw that the smaller (companion) galaxies in the Virgo Cluster had systematically higher redshifts than the larger, older galaxies. We can test and extend our earlier conclusions about age-redshift relations by trying to understand how the Virgo Cluster is structured and how it is evolving. One of the most informative analyses is by C. Kotanyi, and is shown here in Figure 5-1. The upper left panel shows the distribution of the giant ellipticals. M49 is the brightest at the center of the cluster, and there is a line of bright ellipticals above it.

The most interesting panels are at the upper and lower right, which show the spiral galaxies and the radio galaxies respectively. The spiral galaxies are conspicuous for their bright, young stars and earlier integrated spectra. The radio galaxies are emitting radiation due to the rapid motion of charged particles, an activity which would generally be expected to die down with age. As a result, the configuration of galaxies in these two panels looks very much like a giant spiral, with the younger objects out toward the ends of the arms.

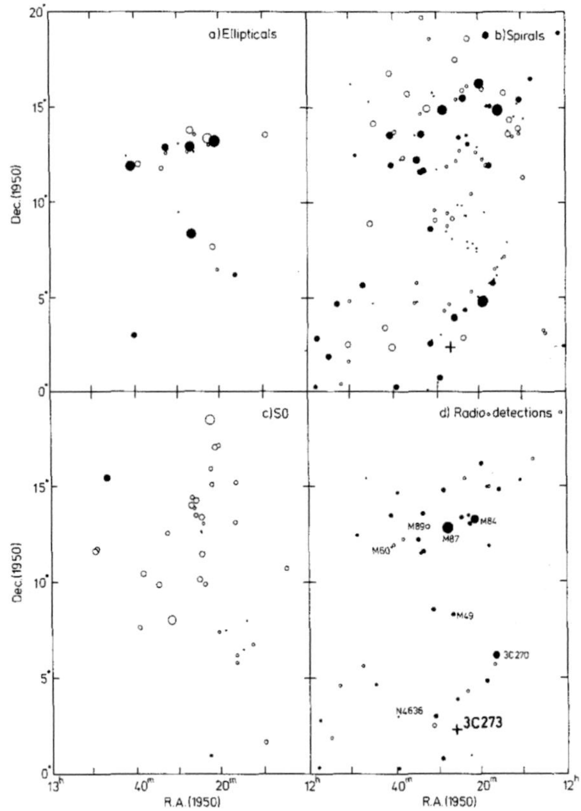

Fig. 5-1. Radio galaxies in the Virgo cluster are indicated by filled circles. Symbol size increases with optical luminosity in a), b) and c). Symbol size increases with radio brightness in d). (From Kotanyi, 1981).

The position of 3C273 had to be added as a plus sign in Figure 5-1d because, even though it was the most conspicuous quasar in the sky, astronomers insisted that its redshift placed it at 54 times greater distance in the background.

A Model for the Virgo Cluster

The conventional explanation for spiral galaxies is that they are in equilibrium rotation, and that a spiral wave runs around the disk condensing the gas and forming new stars. This model is necessary in order to keep material spiral arms, as they appear to be, from winding up in circle after a few rotations (only a few percent of the conventional age of the galaxy). This solution was proposed by a mathematician, C.C. Lin, and quickly became fashionable. There are, however, a number of observational arguments against this model (*IEEE Transactions on Plasma Science*, Vol. PS-14, Dec. 1986, p.748). The alternative model is that material is ejected outward in opposite directions from the nucleus of the spiral, and either the ejection rotates, or differential rotation in the disk draws the ejection track into a spiral form.

The brightest galaxy at the center of the Virgo Cluster, M49, is active. Its activity is is attested to by its rather strong radio and X-ray emission. If it had ejected some material roughly north-south in the past, and then had continued to eject as it rotated

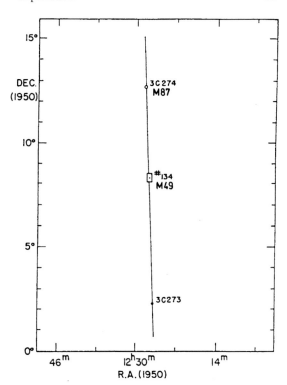

Fig. 5-2. One of the brightest radio galaxies in the sky, 3C274 (M87), and the brightest radio quasar in the sky, 3C273, are paired across the brightest galaxy in the center of the Virgo Cluster, M49 (Arp #134). Alignment is within accuracy of measurement.

counter-clockwise for about 1/8 of a turn, we would expect a spiral pattern roughly like that observed for the spirals and radio detections in Figure 5-1.

What could have been ejected from the central galaxy of the Virgo Cluster? I had stumbled across the answer in 1966 while finishing *The Atlas of Peculiar Galaxies*. Figure 5-2 shows that, aligned directly across M49 (#134 in the *Atlas*), roughly north-south, are two of the brightest radio sources in the sky, 3C273 and 3C274 (M87). The improbability of this being a chance association was computed in 1967 to be about one in a million. Perhaps even more convincing was the common-sense question: Is it significant that the brightest radio quasar in the sky falls in the dominant cluster in the sky—and forms a pair with the brightest radio galaxy in the cluster, almost exactly aligned across the brightest galaxy in the center of the cluster? This result was published in *Science* in 1966 and *Astrophysical Journal* in 1967. It is incomprehensible to me how the field could have gone on believing quasars were at their redshift distances after even this one single result. More than 30 years ago astronomy took a gamble, against odds of a million to one, that this observation was an accident.

Lines of Galaxies along Ejection Directions

In 1968, after the brightest radio galaxies in the sky had been identified, I just looked at the smaller galaxies in their neighborhood and saw them conspicuously aligned across the central galaxy. In most of the cases this alignment coincided with the

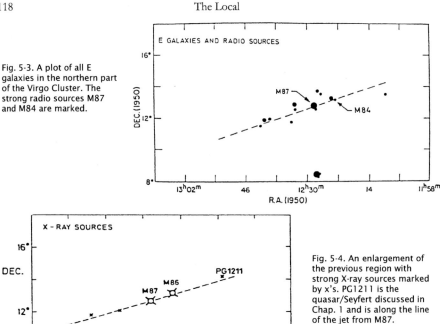

Fig. 5-3. A plot of all E galaxies in the northern part of the Virgo Cluster. The strong radio sources M87 and M84 are marked.

Fig. 5-4. An enlargement of the previous region with strong X-ray sources marked by x's. PG1211 is the quasar/Seyfert discussed in Chap. 1 and is along the line of the jet from M87.

direction of ejection of radio material from the central galaxy (*Pub. Astr. Soc. Pacific* 80, 129, 1968). In a personal letter to me, de Vaucouleurs said there was something very significant here, but he never could make the obvious step that the galaxies in the line had an ejection origin from the large radio galaxy.

In Figure 5-3, the line of E galaxies through M87 (also known as 3C274 and Virgo A) is reproduced from the 1968 paper. It shows how the smaller E galaxies align along the famous blue jet discovered in 1918. In all this time, however, the obvious inference was never considered: that the blue knots in this jet are connected with new, emerging protogalaxies. Figure 5-4 shows the distribution of brightest X-ray sources in this region of the Virgo Cluster. The same line is marked by active X-ray objects! Interestingly, it turns out that the bright X-ray source at the WNW end of the line of the jet is none other than the bright quasar PG1211+143 discussed in Chapter 1. As noted there, the object is non-operationally defined as a quasar, though it is really more like a high-redshift Seyfert galaxy (z = .085). And like so many Seyferts discussed in that Chapter, it is ejecting quasars of higher redshift (z = 1.02 and 1.28). Thus, we have M87 ejecting a Seyfert, which in turn is ejecting quasars. We see a hierarchy of generation, the younger objects having increasingly higher redshifts.

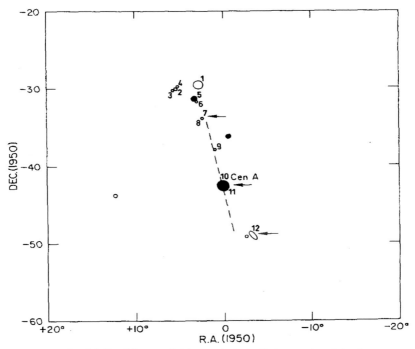

Fig. 5-5. Six of the seven brightest galaxies in this large region of the sky
fall across the giant radio galaxy Centaurus A. Arrows mark the brightest
radio galaxies in the region. Numbers refer to various Seyfert and active
galaxies which fall along this line and which are identified in the text.
Radio and X-ray jets in the center of Cen A are slightly rotated from this
direction.

Returning to the structure of the Virgo Cluster, it is very important to note that in
Figure 5-1 the spirals form a hollow oval just around the line of E galaxies. Again the
experienced cataloguer of the Virgo Cluster, de Vaucouleurs, noted this, but generally
assigned the spirals to a separate cluster *in back* of the Virgo Cluster because they had
systematically higher redshifts. We now know (some of us anyway) that the spirals have
higher redshifts because they are younger. And what else could the hollow oval, mean
but that they had been ejected as protogalaxies from the line of older E galaxies?
Apparently we see at least three generations of galaxies in the Virgo Cluster, and if the
few negative-redshift galaxies are older still, then perhaps we have even more
generations.

The Centaurus A Line

One other giant radio galaxy, the nearest example to us, should serve to clinch this
picture because it shows almost the same structure as Virgo A. Figure 5-5 shows a large
area on the sky of 40 × 40 degrees. Six of the seven brightest galaxies fall along a line
centred on the radio galaxy, Cen A, which is the brightest. The strongest radio sources
(marked by arrows) fall along this same line. Now there is a well-marked radio jet in the
interior of Cen A (NGC5128) and, as in Vir A, a strong X-ray jet coincident with the

radio jet (see *Quasars, Redshifts and Controversies*, p.139). That same reference shows there is a continuous rotation of about 1/8 turn from the direction of the outer line of galaxies and filaments to the direction of the inner jets. This is just about the inferred rotation of the ejection from M49 at the center of the Virgo cluster.

The redshifts of the companion galaxies along this line vary from about the same as the central galaxy to about 4000 km/sec higher. But as Table 5-1 shows, there are a number of Seyfert galaxies along this line which have even higher redshift. The quintessential example is IC4329A (No. 3 in Figure 5-5) which is a very strong X-ray source. This Seyfert is a companion to IC4329, a brighter E galaxy with 560 km/sec lower redshift. IC4329A has identified quasars and candidates around it awaiting redshift measurement. Here we may have a replication of the ejection by M87 of the Seyfert PG1211+143, which in turn is ejecting a pair of X-ray quasars.

In Cen A, we again have three generations of galaxies with increasing intrinsic redshifts as they get younger. There are not any negative redshift galaxies in the Centaurus region, however, and the group seems to have a higher percentage of young and high-redshift objects than Virgo. Therefore, we might consider the proposal that Cen A was ejected as part of the evolution of the Virgo cluster. Cen A is close to the plane of the Local Supercluster and only 53 degrees south in Supergalactic Longitude. Our own Local Group of Galaxies, as well as other nearby bright galaxies, lie very close to the same plane and might well may be associated with Cen A. They might share a common origin from the center of the Local Supercluster, which is the Virgo Cluster.

The Center of the Virgo Cluster

If the structure of the Virgo Cluster is mostly determined by ejection from the central galaxy, M49, we might wonder whether it is running down or is in a quiescent phase. However, while it is true that M49 is not as strong a radio or X-ray source as M87, it is still true that M49 is still a fairly active galaxy in these wavelengths and some

Table 5-1 Identification of Active Galaxies in the Cen A line (Fig. 5-5)

No.	Object	Redshift	App. Mag.	Class	Comments
1	M83	.0009	B_T =8.51	SBc	Strong radio source
2	IC 4329	.014	B =12.60	SO	Weak X-ray
3	IC 4329A	.016	V =13.66		Strong X-ray, Seyfert 1
4	IRAS 13454-2956	.130	V =17.71		Radio source, Seyfert
5	NGC 5253	.0005	B_T =11.11	HII	Star burst, radio, X-ray
6	MS 13351-3128	.082	V =19		X-ray, Seyfert 1
7	IC 4296	.011	B_T =11.6	EO	Double-lobed radio galaxy
8	MCG-06.30.015	.008	V =13.61		Radio galaxy, Seyfert 1
9	Tol 1326-379	.029	V =15.02		Radio, Seyfert 3
10	CENA (NGC 5128)	.0008	B_T =7.89	Epec	Radio, X-ray, jets
11	NGC 5090	.009	B_T =12.6	E2	Radio, X-ray, Comp[n].
12	NGC 4945	.0009	B_T =9.6	Sc	Radio, X-ray, Seyfert

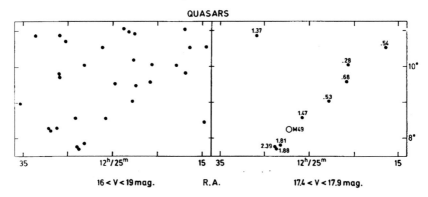

Fig. 5-6. Catalogued quasars in the central region of the Virgo Cluster. On the left all apparent magnitudes. On the right a half magnitude interval of bright quasars. the line of quasars coming from the central M49 contains some close matches in redshift (written next to the symbol) and also quasars close to the quantized values discussed in Chapter 8.

continuing activity might be expected. It is to be noted in Figure 5-1 that M49 appears to have its small retinue of spirals close to the NW.

If we look for quasars, the youngest objects we can identify, we find the intriguing situation pictured in Figure 5-6. The left-hand panel shows a plot of all the catalogued quasars in the area. Just as most experts would expect, there is a random-appearing group of points. But if we plot only the brighter quasars in the ½ magnitude interval between 17.4 and 17.9 mag., magically there appears a line of quasars emerging from M49! The apparent magnitudes turn out to be an indicator of distance. We will continue to see them as useful in this respect, particularly considering that the redshifts are such bad indicators of distance. The redshifts in this line do have an intriguingly close correspondence to the preferred redshift peaks of z = .30, .60, .96, 1.41 and 1.96 (of which we will have more to say later). Also there are some numerically very close pairs of redshifts.

One should also look in the opposite direction from the long quasar line. The quasars emerge from M49 in a NW direction between position angle p.a. = 310 – 320 deg. But about 180 deg. Opposite, at p.a. = 140 deg., there is a disturbed dwarf galaxy, UGC7636. (This is the clue that caused me to put M49 in *The Atlas of Peculiar Galaxies* as #134). Midway between M49 and the dwarf is a hydrogen cloud which Jimmy Irwin and Craig Sarazin argue is so massive that it must be mostly in the form of molecules. I would infer that something must have been flung out or entrained in a direction opposite the ejection of the quasars, as in the typical pairing patterns we have seen throughout this book.

It is also fascinating to note that the redshifts of the aligned quasars tend to decrease as they increase their distance from M49. This is just the behavior shown from the composite of all quasar associations in Figure 9-3, and so dramatically confirmed by the single Seyfert in Figure 9-7.

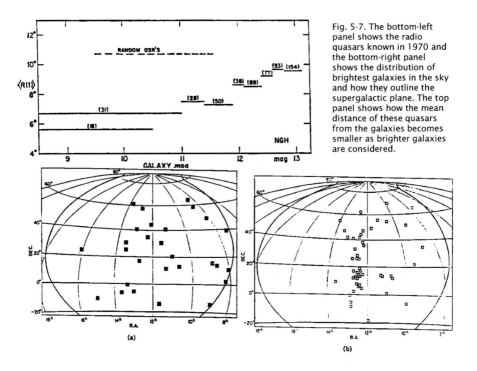

Fig. 5-7. The bottom-left panel shows the radio quasars known in 1970 and the bottom-right panel shows the distribution of brightest galaxies in the sky and how they outline the supergalactic plane. The top panel shows how the mean distance of these quasars from the galaxies becomes smaller as brighter galaxies are considered.

Quasars in the Virgo Cluster

These quasars are not supposed to be there, which makes this is a very delicate subject. But 3C273, the brightest quasar in the sky, was shown to belong in the Virgo Cluster in 1966, only three years after the discovery of quasars (Figure 5-2). In 1970, it was shown that in the Northern Hemisphere the brightest radio quasars were associated with the brightest galaxies—which naturally define the Virgo Cluster (Figure 5-7). Then

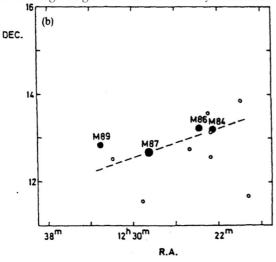

Fig. 5-8. A plot of bright quasars found by X.T. He in the northern part of the Virgo cluster on objective prism plates. Bright galaxies are indicated by filled circles. The association of the quasars with the M87 line galaxies have about one chance in 10,000 of being accidental.

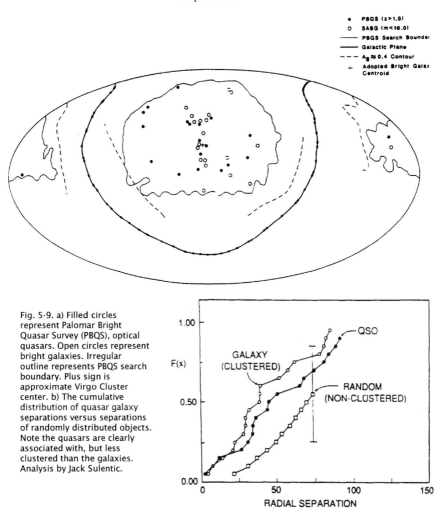

Fig. 5-9. a) Filled circles represent Palomar Bright Quasar Survey (PBQS), optical quasars. Open circles represent bright galaxies. Irregular outline represents PBQS search boundary. Plus sign is approximate Virgo Cluster center. b) The cumulative distribution of quasar galaxy separations versus separations of randomly distributed objects. Note the quasars are clearly associated with, but less clustered than the galaxies. Analysis by Jack Sulentic.

an objective prism search of the Virgo Cluster turned up new quasars, the brightest of which were also associated at a probability level of about one in ten thousand (Figure 5-8). Next came a survey over the northern sky of all bright quasars picked by their ultraviolet colors. This survey was performed by two stalwart believers in "cosmologically distant" quasars, Maarten Schmidt and Richard Green. They failed to notice that their bright quasars were unmistakably concentrated around the Virgo Cluster. Jack Sulentic performed a careful analysis in 1988, and the results are shown in Figure 5-9.

This is primary observational data—simply catalogued positions of quasars—just photons as a function of x and y. And yet it seems to have made no impression on most astronomers who insist on believing that quasars are evenly spread out in the far reaches of the universe.

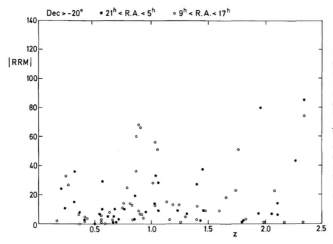

Fig. 5-10. The Faraday (magnetic field) rotation of a number of quasars measured by Kronberg and Perry. Open circles represent quasars in the north galactic hemisphere and represent a jump in polarization in the Virgo Cluster direction at about $z = 1$. This ruins the expectation that the higher redshift quasars would show the most magnetic field rotation because they are the most distant.

Magnetic Fields in the Local Supercluster

Yet another proof of quasars embedded in the Local Supercluster came in 1988. At the same time, it furnished valuable information about the intergalactic medium in the cluster.

> **Faraday Rotation**: The polarized component of a quasar's radio emission will be rotated as it passes through a magnetized plasma. The rotation of the plane of vibration of the photons is called Faraday rotation, and is proportional to the amount of plasma traversed.

Since quasars were supposed to be the most distant objects known, it was of interest to see whether their polarized photons showed any evidence of having traveled through a magnetized plasma on their voyage to us through extragalactic space. Japanese astronomers led by Y. Sofue showed in 1968 that this was indeed true. After careful correction for Faraday rotation due to passing through various paths in our own galaxy, Philip Kronberg and Judith Perry produced a list of 115, of which a subset of 92 are plotted here in Figure 5-10. At first it was claimed that the mean absolute value of the Faraday rotation increased with the redshift of the quasar. That led to the exciting conclusion that the distances to quasars could be measured by their mean Faraday rotation. But then disaster! The mean rotation for quasars around $z = 2$, instead of being twice that of quasars around $z = 1$, was only about $1/3$! But what was a disaster for the conventional redshift distance was just what local quasars required.

Examining Figure 5-10 reveals that the major feature is a large Faraday rotation for quasars of just about $z = 1$. That fits exactly the findings in *Quasars, Redshifts and Controversies*, p.67, where it is shown that around $z = 1$ the quasars have their highest luminosity and therefore can be seen at the greatest distance. (The high rotation for a few of the $z = 2$ or greater quasars is probably due to dense dust/plasma cocoons

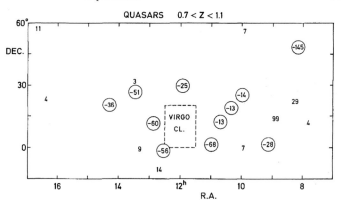

Fig. 5-11. The Faraday rotations for the group of quasars around z = 1 in the direction of the Virgo Cluster show large negative values indicating there is a directional magnetic field in the Cluster.

around these proto-objects.) So the Faraday rotations *can* measure distance—they just confirm unfashionable distances.

Since we can see the z = 1 quasars the furthest, and they are associated with galaxies at distances comparable to Virgo Cluster galaxies, it would be interesting to see if the z = 1 peak comes in any particular area of the sky. Amazing! The open circles in Figure 5-10 show they come from the direction of the Local Supercluster. The editor of *Nature* believed the referees when they said I had cooked the data, and the observers lost interest in the subject when it did not give the expected answers. But actually, there was quite a bit more in the data than the distances.

Figure 5-11 shows the quasars of z between .7 and 1.1 plotted over the region of the north galactic hemisphere. One can readily see their concentration around the Virgo Cluster. Moreover, the circles enclose negative values of Faraday rotation, and it is evident that the Virgo quasars show predominantly large, negative rotations. This is exciting because it means that the magnetic field which is causing the rotation is not alternately one way and then the other, but is dominantly in one direction. We have discovered a systematically oriented magnetic field in the Supercluster center! It is even possible to estimate the field strength as $B = 3 \times 10^{-7}$ gauss or greater (see *Phys. Lett. A* 129,135). This is only a little less than magnetic fields measured in galaxies and over distances more than 2 orders of magnitude greater (volumes a hundred million times larger). This finding will have particular interest for us when we shortly consider the extended, high-energy radiation in gamma and cosmic rays which is coming to us from the center of the Local Supercluster. It is apparent that the high-energy quasars which are embedded in the center of the Supercluster are injecting into a magnetized intercluster medium.

Further Evidence on the Distance of 3C273

In 1989, Riccardo Giovanelli and Martha Haynes announced the discovery of a very peculiar hydrogen cloud in the Virgo Cluster. Like everyone else, I had heard rumors about it, but Geoff Burbidge telephoned me and told me it was only 3/4 degree from 3C273. I was, therefore, very interested to attend a colloquium given by Riccardo at ESO shortly thereafter. When he flashed the map of the cloud on the screen, I

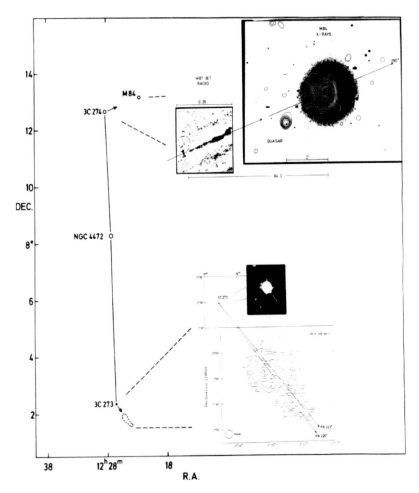

Fig. 5-12. The elongated hydrogen cloud at a redshift near that of the Virgo Cluster is aligned within 3 deg. of the jet from 3C273. On the other side of the central galaxy in the Virgo Cluster we find 3C274, which has a similar size jet pointing to the radio, X-ray galaxy M84, which has swept back X-ray contours indicating that the galaxy is moving out along the line of the jet.

gasped. The cloud was long and narrow and pointed directly back to 3C273! I pointed this out, and there was a moment of silence. Then another member of the audience rather sarcastically asked, "Where does the jet in 3C273 point?" "It points right down the line of the cloud", I replied. There was an even longer moment of silence.

I hurried back to my office and plotted the position of 3C273 and the jet on a Xerox of the cloud that Riccardo gave me. The line of the 3C273 jet lay only 3 degrees off the axis of the cloud, even closer than I had visualized it. The chance of the cloud accidentally pointing back to the position of the quasar times the improbability of the jet pointing down this same line was negligible and, of course, these were both unique objects in the sky. But the immediate importance of this result lay in the fact that the redshift of the cloud was z = 1275 km/sec, about the redshift of M87 and, therefore

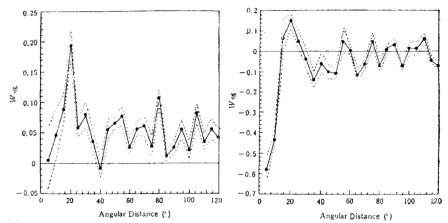

Fig. 5-13. The angular correlation function between quasars and the member galaxies of the Virgo Cluster. On the right is the correlation function between the quasars and non-members in the same area of the Virgo Cluster. Analysis by the Chinese astronomers Zhu and Chu.

(everyone agreed), a member of the Virgo Cluster. *If the cloud was connected with 3C273, that meant 3C273 was in the Virgo Cluster.*

Figure 5-12 illustrates the symmetries involved in the Virgo Cluster. First of all, there is the symmetry of 3C273 and 3C274, so close in the sky that they have successive catalogue numbers—and aligned across M49 (NGC4472). Then they both have the most conspicuous optical jets of any objects in the sky, both almost exactly 20 arc sec long. Along the line of the M87 jet is a radio galaxy with swept back X-ray isophotes indicating motion out along the line of the jet. In 3C273, out along the line of its jet, is the hydrogen cloud like the track of something passing through. The differences are interesting also. There are more, older, lower redshift objects on the M87 side of the cluster, while on the 3C273 side the objects have higher energy spectra and higher redshift, as if this side represented a retarded version of the M87 wing. There are also small radio sources and X-ray sources obviously associated with the jet of 3C273, which should be systematically investigated.

Geoff and I published the material on 3C273, its jet and the cloud in *Astrophysical Journal Letters,* and I spoke my mind on the subject in Patrick Moore's popular magazine *Astronomy Now.* But basically, that ended the matter. It was not ever considered in conjunction with all the other evidence that is gathered together here on the Local Supercluster. One major point of the present book is to try to make it impossible to ignore the enormous amount of mutually supporting, significant evidence which all points to the same conclusion.

The Latest Statistical Association of Quasars with Virgo

In 1995, the Chinese astronomers Xinfen Zhu and Yaoquan Chu analyzed the Large Bright Quasar Survey in the region of the Virgo Cluster. Figure 5-13, shown here, is taken from their work. It illustrates beyond doubt that the 178 quasars clearly fall

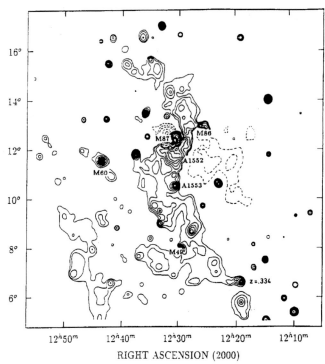

Fig. 5-14. X-ray map of the Virgo Cluster from the ROSAT survey by Böhringer, Briel, Schwarz, Voges, Hartner and Trümper. I have labeled the redshift of the strong X-ray source below M49. See Fig. 6-1a for further identifications.

RIGHT ASCENSION (2000)

closer to the cluster galaxies than they would at random, but are not correlated with the background galaxies. For statistics fans, the confidence level reaches as high as 7.7 sigma in some intervals. It would seem very hard to refute evidence such as that shown in Figure 5-13. These are quasars catalogued by someone else *versus* galaxies catalogued by still another party. The member and non-member galaxies had been determined previously, and exactly the same correlation program was run on both categories of objects.

Because the region is so crowded with galaxies, it is difficult to recognize the individual pairings with quasars more than about 20 arc min separated from their parent galaxy. So an important additional result was obtained by these authors in correlating different apparent-magnitude quasars with different apparent-magnitude cluster galaxies. They found that the more luminous and massive the galaxy, the more numerous the quasars in its vicinity. On the other hand, the brightest quasars were the most separated from their galaxies. Both of these results agree with the analysis of the Seyfert galaxies reported in Chapter 2.

But their results also provided evidence against one of the cherished bastions against quasars being physically associated with nearby galaxies—namely, gravitational lensing. They pointed out there was no association at very small separations, where lensing is most favored. Of course, the wide separation of the bright quasars is much too large for gravitational lensing. These two astronomers have demonstrated enormous skill and integrity in dealing with not only the correlations present in quasars, but also the periodic nature of their redshifts. I will leave it to the reader to guess why the version of their paper which appeared in the European journal *Astronomy and Astrophysics*, unlike the version which appeared in the Chinese *Astronomy and Astrophysics*, did not contain the criticism of gravitational lensing.

X-ray Analysis of the Virgo Cluster

In 1993, a group of X-ray astronomers from the Max-Planck Institut für Extraterrestrische Physik mapped the X-ray photons received from the area of the Virgo Cluster. They published the result in the prestigious journal, *Nature*. It was a very nice paper showing the whole cluster was filled with extended X-ray emission, which they interpreted as a hot gas. One of the diagrams from that paper is shown here as Figure 5-14. I have only added one thing—a label showing that the strong X-ray quasar south west of M49 has a redshift of z = .334. They had identified this object in a previous figure as a quasar, and mentioned in the text that there were some background quasars in the field. But in this figure the quasar just sat there with at least four X-ray isophotes streaming eastward and connecting directly to the central M49.

Everyone who read that article in *Nature* could see that there was a line of X-ray photons connecting this quasar, at 100 times the redshift of the Virgo Cluster, back to the central galaxy of the cluster. What could cause an unrelated background object to behave like this? And all the astronomers involved certainly had heard about the evidence that just these kinds of quasars physically belonged to the Virgo Cluster. What was the most significant aspect of this publication?

I saw this picture in its early stages of preparation and was galvanized by an even more exciting implication. The X-ray emission running down the spine of the cluster was heading more or less southward toward the most famous quasar of all, 3C273, which I knew to be just below the cutoff of the picture frame. I tried to promote a southward extension of the map but got no response. So I asked if I could use the ROSAT Survey observations that went down into this area. The authors were very obliging, and transferred the data into my computer files and gave me a copy of the program they had used to smooth the individual photons into the isophotes of extended X-ray emission. Using the same program as they had on the upper part of the cluster, I produced the southward-extending map shown here in Figure 5-15.

There was one slight difference, in that the X-ray radiation got harder (higher energy) as it went south. So instead of using .4 to 2.4 keV as they had done, I used 1.0 to 2.4 keV energy photons. But I was extremely pleased to see that my iso-intensity contours joined exactly on to theirs. One can see in Figure 5-14 and 15 how the individual features continue across from their map to mine. And of course, there was 3C273 sitting right on the end of a strong, continuous filament connected right back to M49!

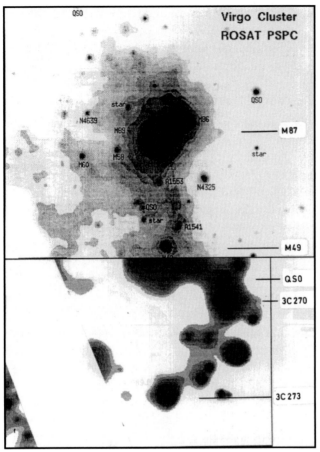

Fig. 5-15. X-ray map going further south in the Virgo Cluster showing the connection of the famous quasar 3C273 back to the X-ray emission in the center. Note how the analysis by Arp of the same Survey data and with the same smoothing algorithm as in the top half of the Figure (derived by the authors of the previous Figure) joins smoothly onto their features.

I don't know quite how to describe the reaction I got to this picture when I showed it to some of my colleagues—perhaps it was like people viewing a grisly automobile accident along the highway. But I had a plan. I knew the upper part of the Virgo X-ray map would get published eventually. I just sat on the map of the lower part until that publication. It took a long time, but when it finally appeared in print, I pounced. In the letter of submission I emphasized to the editor of *Nature*, John Maddox, that since I had used the same ROSAT survey data and the identical reduction program, if their just-published picture was valid, then mine must also be.

What a plan! The editor simply sent me back two referee reports, one more arrogant than the other. I thought surely I could get it published in *Astronomy and Astrophysics*, and revamped the text to make the result less jarring. Three referees turned that down. The kindest one suggested that I take it to the experts who had produced the upper Virgo map so they could explain to me what I had done wrong.

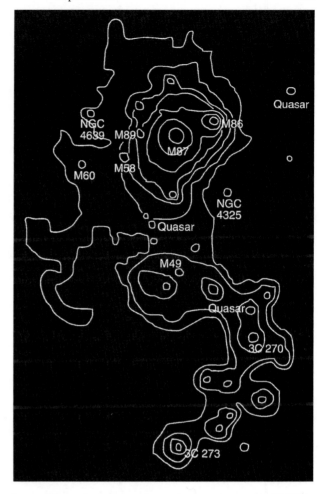

Fig. 5-16. The same map as in Fig. 5-15 above, but now the contour lines are simplified and the important objects have been identified. From an illustration in a news story published by the popular magazine *Sky and Telescope*.

In view of all the other evidence known to show that quasars, and 3C273 in particular, belonged to the Virgo Cluster, I gloomily came to the ironic conclusion that *if you take a highly intelligent person and give them the best possible, elite education, then you will most likely wind up with an academic who is completely impervious to reality.*

My X-ray map was finally published in *Phys. Lett. A* 203, 161 and in the proceedings of the IAU 168 Symposium (in the latter it was presented in the panel discussion that so irritated Martin Rees). From the Symposium, the popular magazine *Sky and Telescope* picked it up and published a simplified contour diagram, which is shown here in Figure 5-16. It is interesting to note that the picture I published in the professional journal emphasized the continuity across the independently measured connecting features so that the viewer could verify that they were real. In the popular journal they assumed that the reader recognized the physical significance of this line of X-ray emission, and instead emphasized the identification of the important objects which were connected together.

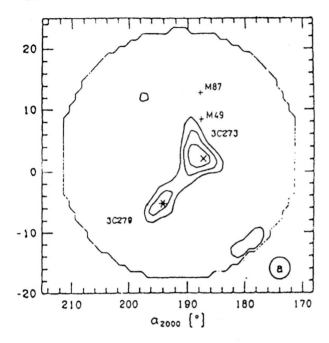

Fig. 5-17. The Virgo Cluster region observed in low energy gamma rays (.7 to 30 MeV range with COMPTEL) by W. Hermsen and 25 collaborators. The blazar 3C279 at z = .538 is shown connected to the quasar 3C273 at z = .158 and then extending back to the center of the Virgo Cluster near M49.

Gamma Rays in the Virgo Cluster

The situation took on added poignancy when observations from the High-energy Gamma Ray Observatory (GRO) started to be processed in an adjoining building. Photons of from thousands to millions of times the X-ray energies were expected to come from only the interiors of the densest, most active extragalactic objects. Yet they appeared spread out over the Virgo Cluster. Worse than that, these extended gamma ray emissions appeared to connect together quasars of different redshift in the Cluster.

A young researcher in the gamma ray division showed me these maps, and we agreed they looked real. The problem was that the senior members of the group were trying to make these extended features go away by elaborate processing techniques, which essentially just consisted of high-cutting the data. Of course, I knew that these observations corroborated and extended the X-ray data in Virgo. It was very painful for us to see this extremely important data withheld.

Naturally, we could only deal with published data, so it was good fortune to find a published map of Virgo in the low energy gamma rays in the proceedings of a meeting held in March 1992. That map is shown here in Figure 5-17. It is clear that 3C273 (z = .158) is linked to 3C279 (z = .538). In the first chapter, Figure 1-12 showed how unrelated sources would meld together, and it is clear that the very elongated iso-contour lines between the two quasars in Figure 5-17 represent a real physical connection.

The quasar 3C279 is violently variable and was, from the Harvard patrol camera records in the early 1900's, once one of the brightest apparent-magnitude quasars in the sky. Like 3C273 then, it is reasonable that it would be associated with the largest aggregate of bright galaxies and radio sources in the sky. It is classified as a "blazar," and is one of the brightest gamma ray sources in the sky.

Of further interest in Figure 5-17 is the extension of these COMPTEL gamma rays upwards toward M49 and M87 in the center and upper part of the Virgo Cluster. To make a preliminary summary of the data, it seems that X-rays in the .1 to 1 keV range extend in either direction from the central M49 to M87 and 3C273. As they near 3C273 they become more conspicuous in the 1 to 2.4 keV range. From 1 to 10 MeV (1000 times more energetic), 3C273 and associated material is most conspicuous. From 10 to 30 MeV, 3C273 begins to drop and 3C279 begins to dominate. This trend continues with the gamma ray telescope on GRO which measures the highest energies, from 100 MeV to several thousand MeV (EGRET). There is a clear progression in the lower energy radiation from the older, low-redshift galaxies to photons of higher energy from the younger high-redshift 3C273, and continuing to even higher energy for the most active, highest-redshift 3C279.

It is important to consult the original reference for the COMPTEL map shown here in Figure 5-17 (it is in *Astronomy and Astrophysics Supp.* 97, 97). Two revisions and four years later, the 26 authors had been reduced to 13, and the observations had been selected and processed in such a way as to show no recognizable connection between the two quasars of different redshift. But fortunately, there was another, more sensitive instrument mounted in the satellite Gamma Ray Observatory, GRO.

Highest-Energy Gamma Rays from EGRET

In the group at MPE processing the observations of gamma rays from 100 to greater than 1000 MeV was the aforementioned Hans-Dieter Radecke. Without knowledge of results in any other energy bands, he pointed out the extended gamma ray emission in the Virgo Cluster, and particularly the gamma-ray extension joining the two quasars, 3C273 and 3C279. There were two reasons why this news provoked great agitation and disapproval. First of all, it placed two quasars in the cluster which officially were supposed to be unrelated to, and at enormous distances behind, the cluster. Secondly, it required these very high-energy photons to come not from dense, energetic cores of cosmic objects, but to radiate from extended, presumably low-density regions.

The researchers were expecting point sources, and when they saw radiation obviously spilling over into extended regions they immediately became apprehensive that their instrument had inadequate resolution and/or uneven background. Considering the cost of the project and future funding, that could understandably cause quite a bit of anxiety. Moreover, there was very little experience at MPE with low surface-brightness astronomical features. The reduction algorithms were designed to detect point sources, and they did that very elegantly. However, I noticed in my own work with the standard data-processing programs that even conspicuous, even slightly extended, sources were often not detected. Since I had done some of the earliest work on low surface-brightness features with the Palomar 48-inch Schmidt telescope, I tried

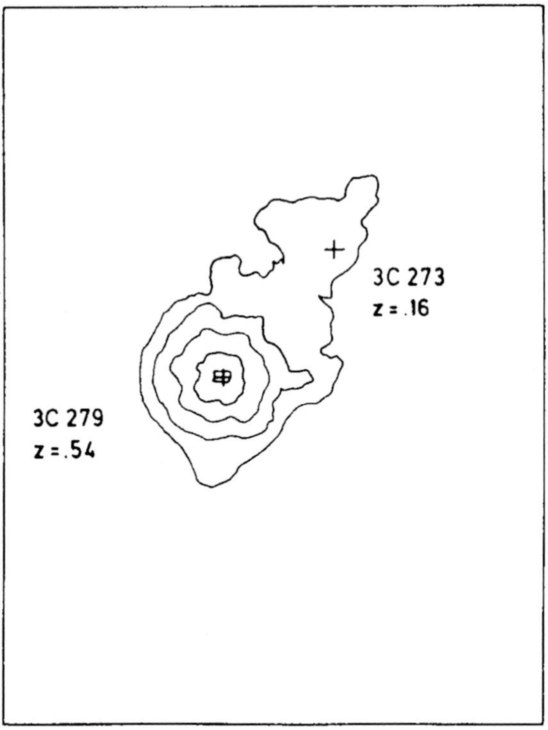

Fig. 5-18. Gamma ray observations (EGRET) of 3C279. Contour lines have been sketched from a color picture in *Sky and Telescope* (December 1992).

3C 273
z = .16

3C 279
z = .54

to convince people of the value of such investigations. But there was only work on objects that were supposed to be extended, like galaxy clusters and supernova remnants. As a result there is a gold mine of data in the archives on galaxies and quasars waiting to be investigated for extended emission.

Radecke encountered strong opposition by some in the group, reluctance and disinterest from others. Nevertheless, he went ahead to try to test the myriad reasons advanced by his colleagues as to why the apparent extended material could not be real. In one test he painstakingly computed the time variability of the photons in the image pixels between 3C273 and 3C279. He was able to show that the bridge stayed constant while the quasars varied strongly in intensity—thus scotching the argument that the bridge was spillover light from the quasars. He even determined a spectrum for the connection, which was manifestly different from that of the quasars.

Of course, anyone could tell the bridge was real just by looking at Radecke's picture, as shown here in Color Plate 5-18. The bridge is long and narrow, and one can easily visualize that it could not be caused by two overlapping circular distributions (as in Figure 1-12). It is also extremely important to note in Plate 5-18 that this high-energy gamma radiation extends from the center of the Virgo Cluster (around Dec. = 8 deg.) down to 3C273, and weakly in the other direction to M87 (around Dec. = 12 deg.). This confirms the conclusion advanced earlier that high-energy objects arise in the center of the cluster and extend outward in opposite direction, but that the extensions to the south acquire increasingly higher energy. (It should be remarked that the observation pictured in Plate 5-18 was taken when both quasars were in a low state—particularly 3C279 was only 5% of its maximum luminosity in the three observing periods. This fortunate circumstance enables us to see the extended connections particularly well.)

Gamma Ray Games

Of course, there was a desire to announce important scientific results from this satellite telescope, so the detection of some known energetic point sources was publicly released. I remember the jolt I received on reading the December 1992 *Sky and Telescope*. There in false color was the gamma ray map of 3C279. Trailing off to the NW was this extension of gamma rays *of which nothing was said in the text or figure caption*. I immediately made an overlay to verify the position of 3C273 at the end of the bridge. Then I copied off the contours from the color picture and produced Figure 5-18. Even with 3C279 at its bright phase the connection to 3C273 showed very well! Are we to believe that the people who released this picture did not know 3C273 was there?

Sometime later, I organized a one-day colloquium in honor of the simultaneous visit of Fred Hoyle, Jayant Narlikar and Margaret and Geoff Burbidge to the Garching Institutes. I invited the MPE group to present some of their material with the dual idea of airing a little mass creation physics and getting a little discussion going on the gamma rays in Virgo. But they simply presented point source statistics, and I had to show the connections shown here in Figure 5-17 and 18 (Radecke's map in Plate 5-18 was not yet completed). A younger member of the group spoke up from the audience and said "Further observations show those features are not real." There were knowing looks all around and all I could lamely say was: "Well I am only able to show the meager data in the literature so, in view of the importance of the point, I think all data should be made available."

After that I made a point of attending gamma ray talks at the Institutes, where they would always present the picture of 3C279 with the trail of emission going off to the NW and make no comment about it. No one else said anything. I had long ago learned that colloquia were events of intense social pressure, and that comments from the floor which questioned the assumptions of the speaker and were not explainable in a few sentences were neither understood nor welcome.

After the Radecke map shown in Plate 5-18 became available, however, a change took place. I attended a presentation to a large audience by the leader of the Gamma Ray group, and when he projected the usual picture of 3C279 I could tell that the bottom half of his transparency contained the picture where both quasars were at minimum and the connection between them was blatant. "Aha", I thought, "this audience is now going to see the inescapable truth of the matter." But after he was through discussing the top half he quickly whipped the transparency off the projector. I suppose he and I were the only people there who knew what was on the bottom half of that transparency.

With a tremendous effort, Radecke finally finished the enormously detailed analysis of the Virgo Cluster gamma rays, and after many delays the group reluctantly

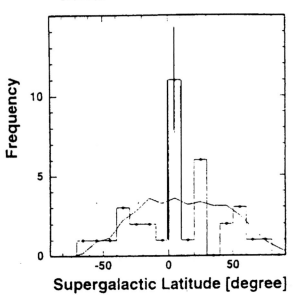

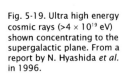

Fig. 5-19. Ultra high energy cosmic rays (>4 × 10¹⁹ eV) shown concentrating to the supergalactic plane. From a report by N. Hyashida *et al.* in 1996.

permitted him to submit it for publication. It was clear by then that he would not be rehired at the imminent end of his contract. After about ten weeks it came back from the appropriate journal with a rejection from a referee in a competing gamma ray group. It was then sent to another journal, and more than two years passed before it finally appeared in *Astrophysics and Space Science*.

After exhausting all possibilities he was about to go on unemployment when the Director of the Institute where I was a guest came to the rescue. For six months he was hired to process the archival X-ray observations around Seyfert galaxies. This was the project described in Chapter 2 under the title "Seyfert Galaxies as Quasar Factories." The project was such a spectacular success in establishing the physical association of high-redshift quasars with the low-redshift active galaxies—thus disproving the whole current basis of extragalactic astronomy—that we felt sure it had rescued his future. I felt this project, plus his knowledge and dedication to the spirit of research, made him the most desirable appointment to a research position of any person I had seen in decades at any level.

But the process of publishing the Seyfert results dragged on and on (we held a little celebratory lunch when the papers were finally accepted). And the publication of his gamma ray results in the Virgo Cluster were still not accepted, much less published. Finally his unemployment money ran out and he had no choice but to change to a career of technical science writing.

Ultra High-energy Cosmic Rays

About the time our hopes of convincing people at the Institutes of the importance of X-rays and gamma rays in the Virgo Cluster were fading , I noticed a development in the Journal *Physics Today*. I don't know how many people are aware of it, but there is a dedicated group of experimental physicists who measure cosmic rays from ground stations in various parts of the world. These cosmic rays are the most energetic radiation we know—some particles, generally changing from iron nuclei to protons at energies of over 10^{19} eV, reach about a hundred million times higher energy than the gamma rays we have been just discussing. They produce air showers of secondary particles which, with some difficulty, are made to yield the arrival direction of the primary particle. The article in *Physics Today* noted that after much data collection, it had been found that the highest energy cosmic rays were coming from the direction of the supergalactic plane and, from the map of events, mostly from the supergalactic center. (One set of measures is pictured in Stanev *et al.* 1995, *Phys. Rev. Lett.* 75, 3056. Another is shown here in Figure 5-19.)

The existence of such high-energy radiation had not gone unnoticed by theoretical physicists and astrophysicists. They doubted that the usual methods of cosmic ray production would give energies that high. Some speculation was forthcoming concerning "primordial matter." This was a shocking suggestion, considering that in the prevailing Big Bang theory all primordial matter had gone "bang" 15 billion years ago. Moreover, the ultra high-energy cosmic rays were calculated not to have an origin at greater distance than about 30 Mpc because of scattering off photons in the cosmic microwave background. That meant they almost had to come from the center of the Local Supercluster, within about 20 Mpc, because this is the only significant concentration of material inside that distance.

Empirically, we had bright quasars, sources of huge amounts of X-rays and gamma rays sitting right in the Virgo Cluster and they were the only possible connection to the ultra high-energy cosmic rays. Yet their redshifts, on the conventional assumptions, ruled them out as residing in the Virgo Cluster. A theory that explained why their redshifts were intrinsically high because of recent origin from energetic, primordial matter in the Cluster would seem to solve everybody's problems.

Will a Theory Help?

In 1993, Jayant Narlikar and I had published a paper outlining how newly created matter would have a high redshift, and demonstrated how to account quantitatively for quasar and galaxy redshifts as a function of their age. (See Chaps 4 and 9 for further discussion). In this Chapter we have spent some time giving the evidence that the powerful quasars and radio galaxies are all young objects taking part in the evolution of the Virgo Cluster. But there had been no specification of how, *i.e.* in what form the low-mass matter was created. Was it electrons and positrons? Hydrogen atoms?

Also in 1993, however, Fred Hoyle, Geoff Burbidge and Jayant Narlikar introduced the quasi steady state cosmology (QSSC). There they created the matter in the form of Planck particles. The mass of the present day Planck particle is about 10^{19} GeV/c^2. In the short time scale of about 10^{-43} seconds the particle is unstable and decays into baryons and mesons. If each particle, for example, decays into 10^9 particles, then each will have energies of around that of the observed UHCR.

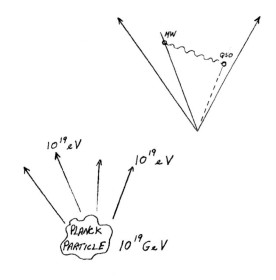

Fig. 5-20. Sketch of matter creation. Top shows a QSO newly created inside the light cone as viewed from our Milky Way galaxy. Middle shows a Planck particle born at zero mass breaking up into particles of energy comparable to observed ultra high energy cosmic rays. Bottom indicates evolution of the Planck masses from the quantum gravity era through Grand Unified Theory to present Electro Weak era.

I had long urged Jayant to create these Planck particles at zero mass and let them grow with time. That would solve the energy and redshift problem simultaneously. Finally, we got together over lunch in Sri Lanka where we had gone to a U.N. sponsored dedication of a new telescope in the Arthur C. Clarke Science Center. The scene in the movie 2001—where the ape threw a thigh bone he was using as a tool up in the air, and it turned lazily over and over until it suddenly transformed into a space station orbiting the earth to the strains of a Strauss waltz—that scene for me was an unforgettable, emotional encapsulation of the potential of humankind. When the author of 2001 later wrote me an enthusiastic letter about my iconoclastic book *Quasars, Redshifts and Controversies* I was thrilled beyond expression.

In spite of my unease at the Tamil rebellion in Sri Lanka at the time, I could not resist flying to Colombo to talk to Arthur Clarke, and at the same time discuss Planck particles and redshifts with Jayant Narlikar. As the lunch was ending we found a piece of paper and he sketched out, simply enough so I could understand it, the diagram which I have copied here in Figure 5-20. This simply illustrates that a single Planck particle has enough energy so that its breakup could supply enormous numbers of ultra high-energy cosmic rays. The evolution path at the bottom simply specifies that in the language of particle physics, the Planck particle is created in the Quantum Gravity era, rapidly breaks up and goes through the symmetry breaking of the Grand Unified Theory, and the resultant individuated particles then proceed into the Electro Weak era.

How this might be applied to activity in the center of the Local Supercluster can be outlined as follows: Newly created material in the centers of active galaxies and quasars is ejected outward in the ubiquitous jets and pairs which are observed. The material starts out with the velocity of light, but slows down as its rest mass grows. The decay products of the Planck particles (*e.g.*, baryons, electrons *etc.*) will show a similar behavior. The energetic particles will penetrate, and be diffused in, the intergalactic medium of the Cluster. As we have seen earlier, this medium is to some extent magnetized, and will wring *bremsstrahlung* and synchrotron energy out of charged particles, particularly low-mass particles. When particles are sufficiently evolved to form atoms, spectral lines produced in transitions would be redshifted, but probably not much more than the observed range in redshifts of the active galaxies and quasars which currently define the Cluster.

What advantages does this model offer?

- It identifies a range of the strongest high-energy radiation in the sky as coming from the unique aggregate of active objects, the Local Supercluster.
- It attributes the energy to ongoing matter creation, a process which replenishes the unavoidably rapid decay of radiation of such high frequency over a very short (compared to cosmic) time scale.
- Since the newly created matter has initially high intrinsic redshift, the younger, higher energy-density quasars and radio galaxies are allowed to be in their apparent location in the Virgo Cluster, and are naturally identified as the principal contributors to the high-energy radiation in the Local Supercluster.
- The extended distribution of the X-ray, and particularly the gamma ray radiation, in Virgo would be a natural consequence of recurrent injections into the intercluster medium.

A Small Epilogue

My ever-naïve plan to win some attention for the observations with this theoretical suggestion was greeted by remarks like: "Well the gamma radiation looks like background irregularities to me," and "The proposed mechanism for the energy is wildly speculative." (From this I conclude that the suggestion that the high-energy radiation comes out of nothing is less speculative!) The reaction around my own institution was so shocked that it was deemed better, if I had to submit it for publication, for me to use my home address! That was all right with me because I was so grateful for the friendly support and the invaluable office facilities that I genuinely did not want to repay my hosts with embarrassment.

But I had to laugh when I thought about sitting in Arthur Clarke's office/reception in Colombo. I wanted very much to bring him some important, not-yet-known scientific development. Something that would surprise and delight him. So when he asked, "Well, what's new in Astronomy?" I pounced.

"There are extremely high-energy cosmic rays, gamma rays and X-rays coming from the center of the Local Supercluster!" I paused to let that sink in and prepared to issue the astonishing and exciting explanation.

"Oh yes," he waived his hand airily: "....matter creation."

I sat there with my mouth open while he went on happily to other subjects. It has slowly percolated through to me since then that there is a world of difference between the imagination of a good science fiction writer and the average professional scientist.

Chapter 6

CLUSTERS OF GALAXIES

Are there other clusters of galaxies which look like the cluster at the center of our Local Supercluster, the Virgo Cluster? Is the universe populated with more distant examples of such great clusters? Everyone believes there are many—and 4,073 of them are listed in the revised northern and southern *Abell Catalogue*. They were originally catalogued by George Abell, and later Harold Corwin, from wide-field Schmidt telescope photographs and are defined as containing, in addition to the brightest two, at least 30 galaxies within a range of 2 magnitudes. The brightest galaxy in the cluster ranges from a little fainter than those in Virgo to about m = 19th apparent magnitude, and redshifts reach about z = .2. (The m used to describe the brightness of the galaxies in the cluster is that of the tenth brightest).

Galaxy Clusters in Virgo

Figure 6-1a is taken from the X-ray map of the Virgo Cluster which was shown in Figure 5-14 of the previous chapter. More identifications of objects, however, have been added. Of the most interest for this chapter are the four bright Abell Clusters running down the spine of the Virgo cluster. They are shown in Figure 6-1b with only their strongest X-ray contours around them. These four are bright X-ray sources and are the only Abell Clusters identified as such within the framed area. *It seems extremely unlikely that this is a chance arrangement.* Moreover, there are individual features of the clusters which support their association with the Virgo Cluster.

My attention was first drawn to the cluster A1541 because the X-ray contours were extended in a direction away from M49. As Figures 6-1 and 6-2 show, the contours also lie along the outer edge of the narrow opening cone of ejected quasars from M49 which was shown in Figure 5-6. The X-ray contours are rather clear in pointing to M49, but could this galaxy cluster at a redshift of z × c = .09 × 300,000 km/sec = 27,000 km/sec be a member of the Virgo Cluster at z × c = 863 km/sec? When the question is asked in this way, the answer is a surprising "yes" because quasars

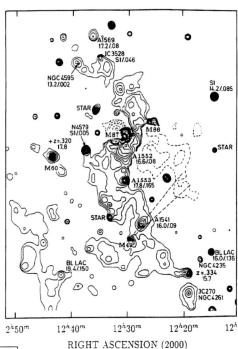

Fig. 6-1a. These are the same X-ray contours in the Virgo Cluster by Böhringer *et al.* as shown in Fig. 5-14. But now more optical objects have been identified. Generally objects are named on the first line and their apparent magnitudes and redshifts are given underneath. Notice the similarity of redshifts among some of the various kinds of objects.

RIGHT ASCENSION (2000)

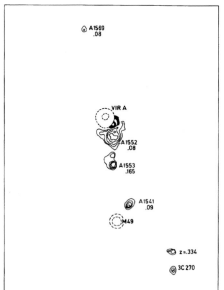

Fig. 6-1b. The four strong X-ray, Abell galaxy clusters in Virgo. Only the stronger X-ray contours are shown from 6-1a.

of much higher redshift were previously shown to be members. But in another sense the question is a shocking impossibility. After all, the most impressive version of the Hubble relation, which reached to galaxies of faintest apparent magnitude and had very small dispersion, was built up by Sandage out of galaxy clusters. Everyone—myself

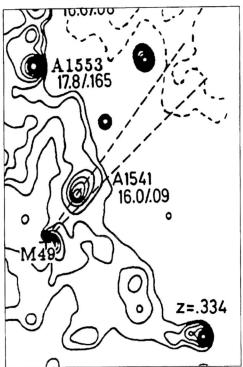

Fig. 6-2. An enlargement from Fig. 6-1a with the dashed lines showing how the line of quasars ejected from M49 relates to the extension of X-ray isophotes around Abell 1541.

included—thinks instinctively of galaxy clusters as galaxies like our own seen at great distances.

The only recourse was to examine the picture further. It turns out that three of the four clusters running along the backbone of Virgo in Figure 6-1b have redshifts of .09, .08 and .08. This would indicate they were related in some way—but if they were distant why would they fall along the central line of the cluster? Particularly A1552 is relatively large in its contours and it seems to emerge from a bright X-ray region on the SW side of the modeled and subtracted image of M87. I know from my plates with the 200-inch telescope that there is a double galaxy about 1 arcmin SW (UGC7652) which is aligned back toward the nucleus of M87 in a configuration very suggestive of ejection. It is of much higher redshift than M87 and probably is connected with the cluster A1552.

In a third cluster along the line, A1569, the contours appear aligned in a direction away from the Seyfert 1 galaxy, IC3528. This appears to be a classic case of the low-redshift (660 km/sec) Sc galaxy, NGC4595, connected by a trail of X-ray emission to a high-redshift Seyfert companion (z = .046). The Seyfert then ejects new X-rays in opposite directions. One ejection coincides with the A1569 cluster at z = .08. It is rewarding to inspect this configuration closely in Figure 6-1a. Note also that the X-ray contours of A1569 point back to the Seyfert!

Finally, the cluster A1553 lies closely on a line between M49 and M87. That line, we know, when extended southward from M49, leads directly to the famous 3C273 which has a redshift of z = .158. When we extend this line northward from M49, it

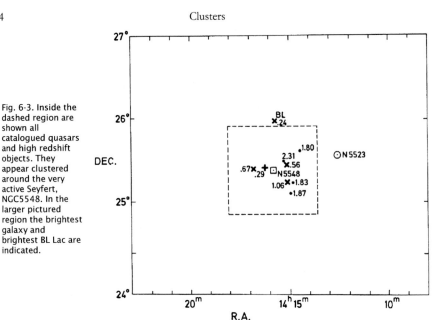

Fig. 6-3. Inside the dashed region are shown all catalogued quasars and high redshift objects. They appear clustered around the very active Seyfert, NGC5548. In the larger pictured region the brightest galaxy and brightest BL Lac are indicated.

leads to A1553 which has a redshift of z = .165. Is this the counter ejection to 3C273? Is M87 instead an earlier counter ejection?

People have sometimes said to me: "If you make extraordinary claims you must have extraordinary proof." Usually they exit smiling smugly and I think dark thoughts like "no proof is extraordinary enough." But if something is true there is always the possibility that you can find some extraordinary proof.

A Connection with the Seyfert NGC5548

At about this same time around 1993, when H.C. Thomas showed me his observations of a bright BL Lac object, I noticed 36 arcmin south of it a strong X-ray source with an apparent jet. This source turned out to be the very active Seyfert galaxy NGC5548. Since the field had been included in the Einstein Laboratory, medium sensitivity survey, it was known that there were three X-ray quasars within about 15 arcmin of z = .56, .67 and 1.06. Also there were catalogued quasars of z = 1.80, 1.83, 1.87 and 2.31 nearby. Clearly, as Figure 6-3 shows, there was a concentration of active objects in this small region of the sky.

But most important for the purposes of this chapter is the apparent X-ray jet emerging in the NE direction from NGC5548. With the help of Hans-Christoph I was able to obtain the IPC frame shown here in Figure 6-4. The "jet" is visibly broadened and centered about 8 arcmin from the nucleus of NGC5548. There appears to be a narrow break in the continuity of the X-ray emission between NGC5548 and the X-ray object. Later Radecke and I were able to obtain a ROSAT exposure from the archives, and this slightly deeper, slightly lower energy range in Figure 6-5 shows in addition to elongation of the X-ray object toward NGC5548, a narrow filament joining the two objects.

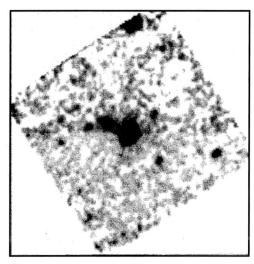

Fig. 6-4. Einstein Laboratory X-ray map of NGC5548 showing strong jet-like emergence which turns out to be a galaxy cluster of redshift z = .29.

The optical identification of this "jet" or elongated X-ray emission is therefore crucial. On the Palomar Schmidt Sky Survey at this position there is, at the extreme limit of the red and blue prints, what appears to be a faint, elongated cluster of galaxies. This seems to be confirmed by the measure of John Stocke *et al.* of a non emission line galaxy at z = .29. (Note the closeness to a major quasar with redshift peak at z = .30—and consider that if you broke up an 18th apparent magnitude quasar into 16 pieces you would have fainter components about m = 21 mag.).

Because of the new evidence that some faint "galaxy clusters" are in fact ejected from active galaxies, the cluster adjoining NGC5548 becomes a crucial case for further investigation. High resolution, deep images and spectra with Hubble Space Telescope or large ground based telescopes would seem to be of the highest priority.

Fig. 6-5. Smoothed X-ray contours from slightly deeper, slightly softer exposure from ROSAT showing elongation of cluster and possible bridge to NGC5548.

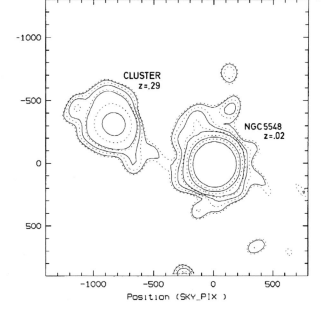

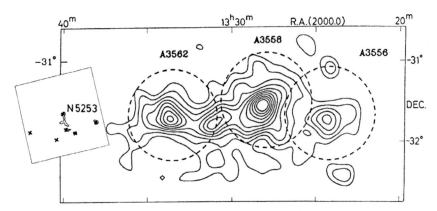

Fig. 6-6. The contour lines outline the optical density of galaxies in three adjoining, rich Abell clusters (after Bardelli *et al.*). The inserted box shows the location of one of the brightest star burst galaxies, NGC5253, and X-ray sources in its field. Two of the three Abell clusters are strong X-ray sources.

Three X-ray Clusters in a Line

In a separate event about this time I happened to notice an article in the *European Southern Observatory* (ESO) *Messenger* which pictured the distribution of galaxies in three adjacent Abell Clusters (Bardelli *et al.* 1993). Figure 6-6 shows that these three clusters define approximately a straight line. But it also happens, as the Figure shows, that a very bright, very active galaxy falls at about the terminus of this line. NGC5253 is a well-known, exceptionally active starburst galaxy. It is also an active X-ray galaxy with a filament coming south and then splitting E and SW with at least three additional X-ray sources leading over into the line of the three Abell Clusters (insert in Figure 6-6).

Two of these three Abell Clusters are bright in X-rays. In a list of X-ray clusters ranked by apparent flux (Lahev *et al.*) these clusters rank no. 26 and 29. About the same distance on the other side of NGC5253, however, lies A3571, an exceptionally bright X-ray cluster. In apparent flux it ranks fifth after such clusters as Virgo, Coma and A2319.

So the situation we are faced with is that three clusters, two of which are strong X-ray sources, form a line which appears to originate near a bright starburst galaxy. This galaxy is itself an X-ray source with a jet, and has X-ray material that appears to extend in the general direction of the line of clusters. In sum, we find the X-ray galaxy NGC5253 near the center of a broken line (angle about 35 deg.) of unusually bright X-ray bright clusters.

It was natural at this point to look at the disposition of all Abell clusters over a wider area around NGC5253 in order to appraise the possibility of the observed configuration occurring by chance.

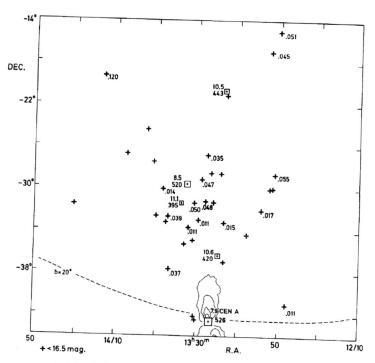

Fig. 6-7. All Abell galaxy clusters with m equal to or brighter than 16.5 apparent magnitude are plotted over this approximately 30×30 degree region of the southern sky. Small boxes identify brightest galaxies with their apparent magnitudes and redshifts in km/sec. Contour lines of radio emission from the giant radio galaxy Centaurus A are shown. The dashed line shows the location of galactic latitude 20'.

Plot of all Bright Clusters North of Cen A

Investigating the cluster population around NGC5253 made it clear that it was necessary to plot cluster positions over a wider and wider area in order to judge the average background density of clusters. Figure 6-7 is the result and shows that just north of the 11.1 mag. NGC5253 lies the 8.5 mag. M83 which has a suggested line of clusters passing through it. (And, in fact, M83 may claim the strong X-ray cluster NE of NGC5253. It is a fascinating game to try to find which objects belong to which bright galaxies in crowded regions.)

Just south of NGC5253 is the active radio galaxy IC4296 which is near some clusters of redshift about z = .011. (IC4296 is below the cut off in apparent magnitude for plotting in Figure 6-7, but other studies have indicated it is a member of the line of galaxies associated with CenA—see Figure 5-5 here and also *Pub. Astr. Soc. Pacific* 80,129,1968 and *Quasars, Redshifts and Controversies*, p142). The situation is that there are so many bright and active galaxies in the CenA line that it is difficult without further study to ascertain which clusters belong to which galaxies. *But the overall result of Figure 6-7 is even more stunningly clear and important—namely that the bright Abell clusters of galaxies in this*

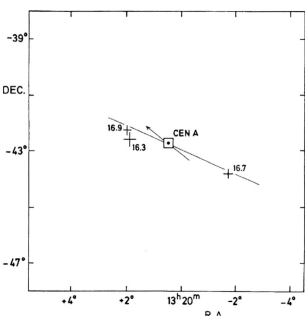

Fig. 6-8. An enlargement
of the region around
CenA showing all Abell
clusters with m = 16.9
or greater. The smaller
arrow indicates the
direction of the strong
radio and X-ray jets in
the interior of CenA.

*large region of the sky are distributed in the same characteristic way as the bright and active galaxies
that belong to CenA!*

Since the higher redshift galaxies in the CenA line are presumed to have originated
in ejection from this giant, active radio galaxy, the Abell clusters so densely surrounding
these second generation galaxies are implied to be third generation ejecta—in various
directions but still relatively close to their galaxies of origin. *CenA, at the bottom of Figure
6-7, with its outer radio isophotes sketched in, resembles a flame with the sparks of galaxies and
clusters rising upward.*

If we ask whether there are any clusters directly associated with CenA, we can
consult the smaller field shown in Figure 6-8. There we see only three catalogued
clusters and all are brighter than m = 16.9 mag. Two lie on one side of CenA and one
on the other side. All three are seen in X-rays in the analysis of a 10 × 10 deg. ROSAT
field (*A&A* 288,738). The direction of the innermost X-ray and radio jets is indicated
by the arrow. *The disposition and alignment of these three closest clusters would seem to unequivocally
establish their primary ejection origin from CenA.*

Two further aspects should be noted about the clusters in the larger region around
CenA: First, the Abell clusters in Figure 6-7 which have measured redshifts fall into two
distinct groups. There are six with .011< z < .017 and seven with .035< z < .055. This
segregation can also be seen in the Hubble diagram of Figure 6-14. Redshift
discretization is typical of galaxies with intrinsic redshifts, is evidence against distance
related redshifts, and is probably due to the episodic nature of matter creation events
(Arp, *Apeiron*, 9-10,18,1991 and Narlikar and Arp, *Astrophysical Journal* 405,51,1993).

Secondly, the most significant concentration of clusters over the huge region in
Figure 6-7 occurs for the brightest Abell clusters, as plotted (*i.e.* those clusters

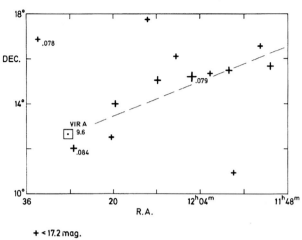

Fig. 6-9. All Abell clusters with m equal to or brighter than 17.2 mag. in a region centered on Virgo A and its jet (direction of dashed line). The clusters of redshift .078 to the NNE and .084 to the SW were identified in Fig. 6-1b as belonging to the Virgo Cluster. Symbol sizes proportional to m brightness.

+ < 17.2 mag.

containing galaxies brighter than m = 16.5 mag.). We will see in the following sections that other active radio galaxies show strong associations with bright Abell clusters. But Cen A is the closest of these systems and, as would be expected, its associated Abell clusters are the brightest and most widely spread.

Abell Clusters Associated with Virgo A (M87)

Proceeding from CenA to the next brightest radio galaxy in the sky, VirA, we encounter another giant E galaxy in which the radio jet, X-ray jet and now an optical jet are all coincident and define a precise ejection axis. Perhaps even more accurately than in CenA, numbers of bright E galaxies fall along these ejection directions from VirA, defining a length on the sky of about 8 deg. as shown in Figures 5-3 and 5-4 in the previous chapter. In Figure 6-9 all catalogued Abell clusters with m brighter than 17.2 mag. are plotted in the area of VirA and the principle extension of its jet direction. It is clear from that diagram that the Abell clusters also define the direction of that same ejection line. Note also that the quasar/Seyfert PG1211+143 discussed in Chapter 2 lies closely along this same line.

The brightest clusters (indicated by the size of the plus symbol) define the line best. Two clusters along the line have measured redshifts of z = .084 and .079 indicating empirically that they are associated even though quite far separated along the line. (The fainter cluster in the NE corner with z = .078 is Abell 1569 associated with objects in the X-ray extension NNE from M87 shown in the map in Figure 6-1.)

Abell Clusters in the region of Perseus A

In 1968 it was shown that essentially all bright radio galaxies had lines of smaller galaxies emerging from them. PerA is another very strong radio galaxy and it is well known that a long straight line of galaxies proceeds almost due west from it. What is not so well known is that this chain of galaxies ends on a very bright, nearby galaxy. Figure 6-10 shows all Abell clusters in the region with m < 17.4 mag. It is clear that there is a

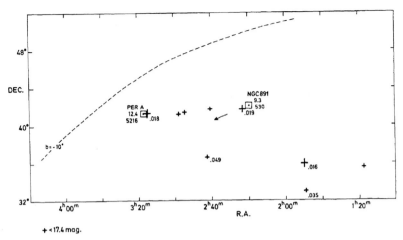

+ <17.4 mag.

Fig. 6-10. Abell clusters with m brighter than 17.4 mag. in region around PerA
and NGC891 (latter at apparent magnitude 9.3 and redshift 530 km/sec.).
Symbols in proportion to cluster m brightness. The Perseus-Pisces filament of
galaxies runs from PerA to NGC891 and then SW as in the following figure.

straight line of clusters connecting PerA with NGC891, the tenth brightest Sb galaxy in
the sky at northern declinations.

NGC891 is well known as a nearly edge-on disk galaxy with, consequently, its
minor axis well defined on the sky. That direction is close to east-west and is indicated
by the arrow in Figure 6-10. The reason this point is noted here is that although
NGC891 is not at present a particularly active galaxy, spirals do intermittently eject
material and the evidence favors ejection along the minor axis. (*E.g.* see Figure 1-10.)

Of course, what we are seeing in Figure 6-10 is the famous Perseus-Pisces chain of
galaxies. This narrow chain stretches about 90 deg. across the sky! The redshifts of its
galaxies characteristically range between 4000 < cz < 6000 km/sec. If this dispersion in
redshift were to be translated into velocity, the filament would spread to more than ten
times its observed width in the supposed age of the galaxies. This, together with its
extraordinary length and circularity centered on the observer, if it is at its redshift
distance, argues for intrinsic redshifts and a much nearer distance for the chain (*Journal
of Astrophysics and Astronomy* (India) 11, 411,1990).

The redshifts of the two bright Abell clusters at either end of the straight line
connecting PerA and NGC891 in Figure 6-10 are typical redshifts for galaxies in the
Perseus-Pisces filament. To the SW of NGC891 is a bright cluster of z = .016, again a
characteristic redshift for the filament. In other words the chain of galaxies
recommences, after a gap, as if the counter ejection from NGC891 had been directed at
a somewhat different angle. This is reminiscent of the possible configuration of clusters
around NGC5253 discussed two sections earlier. In order to fully appreciate the
involvement of NGC891 with the Perseus-Pisces filament, one should also consult the
computer plot of all Zwicky Catalogue galaxies by R. Giovanelli as shown here in Figure
6-11. It can be seen there that both sides of the long filament deviate from their course,
on the west side to point toward, and on the east side to essentially join, the position of

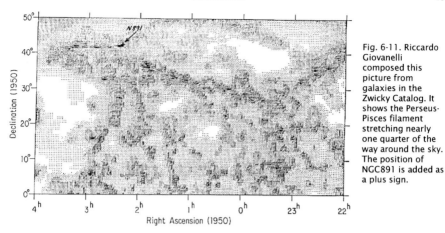

Fig. 6-11. Riccardo Giovanelli composed this picture from galaxies in the Zwicky Catalog. It shows the Perseus-Pisces filament stretching nearly one quarter of the way around the sky. The position of NGC891 is added as a plus sign.

NGC891. It will be seen later in the discussion of galaxy alignments, that most of the bright galaxies in the sky have broken lines of higher redshift galaxies originating from them.

Testing the Ejecting Spiral, NGC1097

This barred spiral is a very bright Seyfert, hot-spot, starburst galaxy with the longest, straightest optical jets known. To test the hypothesis of ejection of clusters the region around NGC1097 was examined in the Abell Catalogue of galaxy clusters. The upper right corner of Figure 6-12 shows that *if the four optical jets in NGC1097 are extended outward, they include within their ejection angles all the Abell clusters in the vicinity brighter than m = 17.0 mag.*

That these directions really represent the ejection of intrinsically high redshift objects is attested to by the color Plate 2-7 and all the studies of NGC1097 referred to in Chapter 2. In the NE direction, just where the optical jets are the strongest, the excess numbers of quasars were the greatest. In the SW direction, where the optical jets were weaker, the excess numbers of quasars were the next greatest. Now it turns out the number of clusters is also correlated with the strength of the jets.

In the process of examining the NGC1097 field, however, I noticed a number of galaxy clusters toward the SE. I had forgotten about the large, nearby Fornax Cluster!

Fornax A—the Twin of Virgo A

As Figure 6-12 shows, the brightest galaxy in this 28x 24 deg. region of the sky is NGC1291 at $B_T = 9.4$ mag. and $cz_o = 738$ km/sec. This is the center of what turns out to be an almost exact duplicate of the Virgo Cluster in roughly the opposite quadrantof the sky! The resemblance is uncanny. For example the brightest galaxy in the center of the Virgo cluster, NGC4472 (M49), is $BT = 9.3$ mag. and $cz_0 = 822$ km/sec, an almost exact match for NGC1291!

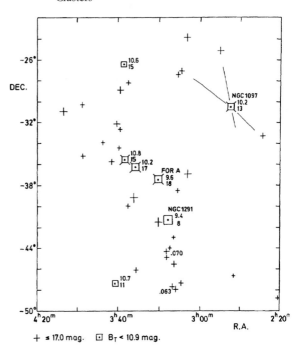

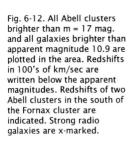

Fig. 6-12. All Abell clusters brighter than m = 17 mag. and all galaxies brighter than apparent magnitude 10.9 are plotted in the area. Redshifts in 100's of km/sec are written below the apparent magnitudes. Redshifts of two Abell clusters in the south of the Fornax cluster are indicated. Strong radio galaxies are x-marked.

In the Fornax Cluster there is a very bright, very strong radio galaxy, ForA, only about 4 deg. away from the central galaxy. In the Virgo Cluster we have VirA (M87) only about 4 deg. away from NGC4472. These are very close matches also (BT = 9.6 mag. and cz_0 = 1713 km/sec for ForA and BT = 9.6 mag. and cz_0 = 1136 km/sec for VirA).

From ForA there are two radio galaxies in a straight line, suggesting an ejection direction. The inner radio jets of ForA are aligned at p.a. = 126 ± 14 deg. But the major axis of the spheroid, as well as the unresolved optical knots which are aligned along it, are at p.a. = 59 ±1 deg. This is the alignment of the two radio galaxies toward ForA, p.a. = 61 ±1 deg., and it would seem to delineate the outer ejection locus of the major galaxy line (as in CenA) of the galaxies aligned with ForA.

But now notice the plus signs in Figure 6-12 which represent the Abell clusters of galaxies. They fall right down the spine of the Fornax Cluster! *This is the same distribution of clusters we found in the Virgo Cluster, and it cannot represent an accidental projection of background objects.* These bright "galaxy clusters" must belong to the nearby great clusters, Virgo and Fornax.

In order to emphasize the amazing similarity of the two great clusters, I have superposed in Figure 6-13 the X-ray outline of the Virgo Cluster from Figure 5-15 on the Fornax Cluster. The scale on the sky is the same, but I have mirror imaged the outline of the Virgo X-rays—*it is as if we were looking at two identical clusters from a point midway between them.* The reader will have to follow the implications further, but I will add that doubling and twinning of cosmic objects seems, in my experience, to be common.

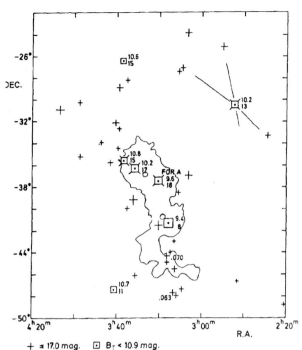

Fig. 6-13. The same diagram of the Fornax Cluster as in Fig. 6-12 but now with the X-ray outline of the Virgo Cluster from Fig. 5-15 drawn in. (The Virgo Cluster outline has been mirror imaged and rotated slightly but is the identical scale). The lower circle represents the position of M49 in the Virgo Cluster and the upper circle represents the position of VirA.

Of course I have tried, from the ROSAT Survey, to map the extended X-ray emission in Fornax as it was done in Virgo. But there are satellite scans which were corrupted as they passed near the Fornax Cluster. So the test of the prediction made in Figure 6-13 will be an exciting possibility only from the next-generation, all-sky X-ray survey instrument.

The Hubble Diagram for Galaxy Clusters

The immediate objection to nearby clusters of galaxies is that the clusters are supposed to be composed of galaxies just like those in our cluster or the Virgo Cluster. How do we know this? Because they have large redshifts, which place them at large distances, and therefore require the galaxies to be luminous and large. But the whole thrust of the evidence so far in this book has been to show that extragalactic redshifts do not generally mean velocity and that younger objects have high intrinsic redshifts, closer distances and lower luminosities. Are many fainter galaxy clusters composed of young objects?

Before we tackle that question, let us test the one observational relation which the clusters must pass if they are at their conventional distances—namely, do they define a Hubble relation? Figure 6-14 shows the 14 clusters with redshifts from the large region north of CenA pictured in Figure 6-7. The best-fitting Hubble relation is shown by the dashed line. *The spread from this line is enormous.* The classical dispersion around a Hubble relation for supposedly distant clusters is only a few tenths of a magnitude. The total

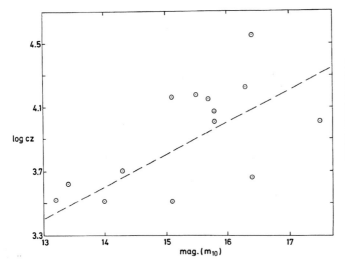

Fig. 6-14. All Abell clusters with measured redshifts from the large area north of CenA shown in Fig. 6-7. The best-fitting Hubble line is shown dashed. The maximum deviation from this line is about 30,000 km/sec.

range in Figure 6-14 at a given redshift, in contrast, is about 4 magnitudes! That would require the luminosity of a characteristic galaxy in a cluster to range over a factor of 40!

Of course the magnitudes which are used in the classical determination of Hubble slopes have a whole series of corrections made to them, whereas the magnitudes listed for the Abell clusters are simply estimates made from Schmidt telescope photographs. The corrections which can be made to these photographic estimates, however, are very small compared to their range of deviation from the Hubble line in Figure 6-14. Therefore we must conclude that the apparent magnitudes of the galaxies in these clusters correlate very poorly with their redshift. Measured in another way, at a little fainter than 16th apparent magnitude in Figure 6-14, the highest-redshift cluster exceeds by about 30,000 km/sec the redshift which a cluster with average-luminosity galaxies should have. The Hubble line is supposed to have a velocity dispersion of about 50 km/sec! Even the most adventuresome observers have only claimed peculiar velocities of clusters between 1000 and 2000 km/sec. The upshot is that although the clusters with fainter galaxies tend to have the higher redshifts, *there is no redshift-apparent magnitude relation for these clusters like that which is claimed to demonstrate a redshift-distance relation.*

In Chapter 9 we will attempt to make a more quantitative analysis of this situation, but now we would like to gather some more examples of this shocking result involving nearby clusters of galaxies.

NGC4319/Mark205—A Very Active System

The disrupted spiral galaxy NGC4319 was first noticed because the bright Seyfert/quasar Markarian205 fell only 40 arc sec away. Of course the chance of this happening by accident was negligible, but conventional astronomers were certain it was an accident because the redshifts of the two objects differed by about 20,000 km/sec. When a luminous connection was discovered between the two, of course a great fight erupted, with the conventional side saying the connection did not exist (see *Quasars,*

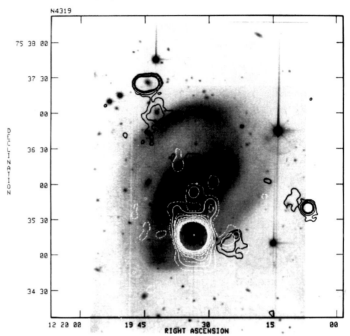

Fig. 6-15. The disrupting galaxy NGC4319 and the Seyfert Markarian 205. The X-ray contours are indicated in white and the radio contours measured by Jack Sulentic are indicated in black. A line of ejected radio objects is seen running NE and SW. Three radio galaxies are identified; the northern ones are seen to be in a galaxy cluster.

Redshifts and Controversies). But the common sense point, which no one ever said much about, was the fact that the galaxy, NGC4319, was obviously in the process of being destroyed. One might argue that the close proximity of the very active Mark205 was responsible—but an even more plausible reason was apparent in the fact that the two spiral arms of NGC4319 were coming off at their roots. There apparently had been a rather violent and recent explosion in the center of the galaxy. This was supported by the fact that most of the expected gas in the galaxy was missing—presumably blown out—and the nucleus missing or inactive.

The arguments that Mark205 has been ejected from NGC4319 have been detailed in my previous book. The account of how Jack Sulentic measured the only good radio map of the system with the Very Large Array in New Mexico and demonstrated the surprising fact that the spiral was also ejecting radio material has also been recounted. But what was not discussed was the fact that these ejected radio lobes were centered on some rather ordinary looking galaxies of considerably higher redshift. We agreed that these looked like radio galaxies in a cluster of galaxies. I said to Jack, "Someday we will have to face the fact that these objects were also ejected out of NGC4319." He just looked appalled and shook his head.

So now has come the time I have to point out that the patches of radio emission in Figure 6-15 lead from the center of NGC4319 out in the NNE direction to two radio lobes unmistakably centered on moderately faint galaxies. The redshift of the stronger radio galaxy is $z = .343$, and there are obvious fainter galaxies which form a cluster. In the counter direction, a radio lobe encircles a fainter galaxy, which has an apparent tail of radio *and* X-ray emission coming off to the SE. The latter represents a significant connection between the classical, double-lobed radio ejections and the younger, higher redshift X-ray objects which were also shown to be ejected in earlier chapters of this book.

Further out along the line of X-ray material which leads from Mark205 to the $z = .464$ quasar is a strong X-ray source marked CL in Figure 1-7 (also visible in Plate 1-7). That source has been identified as a cluster of galaxies at redshift $z = .240$ by John Stocke *et al.* What this means is that we have *two clusters of different redshift situated along lines of ejection* from the Seyfert Mark205 and the interacting, disturbed spiral, NGC4319.

These are just the kinds of cluster galaxies which have been shown in the earlier sections of this chapter to be associated with low-redshift, ejecting galaxies. So NGC4319/Mark205, where there is direct evidence for the ejection of a cluster of galaxies of $z = .343$ and .240 from the active galaxy, represents another example supporting the case of NGC5548. Since NGC4319 has also probably ejected Mark205, which in turn has ejected the quasars shown in Chapter 1, it is perhaps understandable that NGC4319 looks somewhat exhausted!

Other Candidates for Ejected Clusters

No attempt can be made here to be complete because there are so many instances of high-redshift galaxy clusters with evidence for ejection origin from low-redshift, active galaxies. But a few cases will be cited to give a flavor of the kind of evidence which could be followed up with further observations:

- In Figure 1-1 in the first chapter of this book, NGC4258 shows a typical extended X-ray source off the NW end of its main disk (an X-ray cluster of galaxies). That source contains a line of moderately faint galaxies which is in somewhat the same direction as HII regions on that end of the galaxy. (Note that aligned galaxies are a mark of a non-equilibrium configuration, because in their conventional lifetime even small peculiar velocities would disrupt the line.)
- NGC4151 shows a conspicuous straight line of fainter galaxies to the west of its nuclear region in Figure 3-16. This line of galaxies points closely back to the very active companion Seyfert (the ScI galaxy NGC4156). From what we are now seeing, this does not look so much like a coincidence. The five redshifts measured in this line from left to right are: .060, .160, .158, (.16:) and .056 and/or .24. The redshifts in agreement indicate cluster-like association, and the ones in disagreement represent anomalous redshifts because they are obviously part of the same linear feature. (Also note the agreement of the redshifts with previous quantized peaks.)
- X-ray maps of the edge-on galaxies NGC4565 and NGC5907 show X-ray clusters of galaxies near the nucleus which show evidence of elongation back toward the nucleus.
- The X-ray galaxy cluster Abell 85 (see Chapter 8) is only 41 arc min north of NGC217 (MCG 02-02-085). It is 10 to 15 degrees off alignment with the minor axis, and extended toward, NGC217!

- Extensive studies of active galaxies like the starburster NGC253 and NGC3079 (Wolfgang Pietsch *et al.*) show X-ray emission extending out into the halo regions, especially along the minor axis directions. For NGC253 some of this emission is identified with faint galaxy clusters. I would take this as very strong evidence for clusters of small, intrinsically redshifted galaxies to originate by ejection along with other high-redshift material like quasars and BL Lac objects.
- On Palomar and U.K. Schmidt survey plates, I have seen occasional galaxies with streamers of faint galaxies extending away from them.
- In the (ESO) *Messenger* 92,32,1998 there is a definitive picture of quasars and cluster galaxies emerging like a fountain from a larger, lower-redshift galaxy.

Many more examples of galaxy clusters falling provocatively close to low-redshift galaxies could be enumerated but, of course, what should be done is to cross correlate lists of the two kinds of objects by computer analysis and then investigate critical individual cases in detail.

Who Can Consider Drastic Change?

In 1993 I was very excited because this was a staggering change in concept, but one that seemed inescapably required by the observations—after all, look at Figure 6-1b, 6-2, Figure 6-7 and Figure 6-13 just by themselves! But about that time the best scientists I had personally worked with, and valued personal friends, visited the Garching institutes—Fred Hoyle, Margaret and Geoffrey Burbidge and Jayant Narlikar. It was nice to have them there because I felt it bolstered my unpaid visitor status at the Max-Planck Institut für Astrophysik. With great zest I laid out before them what I had discovered about the galaxy clusters. They were horrified!

Geoff said that the redshift-apparent magnitude relation, which was accepted for clusters of galaxies (their Hubble relation) meant that they had to be at their redshift distances. Fred said that my embracing such an obviously crazy result would undermine the credibility of our attack on the Big Bang. He was visibly angry. How could I do this to the person who had thrilled me beyond expression by coming up from the Cal Tech campus to my office at Santa Barbara St. to see my original observations in the late 1960's?

Perhaps I was hoping for some support and advice on strategy—but it was clear that the people I admired the most thought I was ridiculous. Through my disappointment, I had to admit they were completely right, that the result and everyone remotely connected with it would be ridiculed mercilessly.

Struggling with feelings of shame and apprehension at the same time, I felt the results were correct, and I had to think of a way to communicate them. My first thought was to send them to the *Indian Journal of Astronomy and Astrophysics*. It would be almost certain that they would not be read there by any of the establishment bulldozers, the results would be referenceable and at least would not be lost. That reminded me that a few years before I had published in that journal another unbelievable result—alignments of faint high redshift galaxies across almost all the closest, bright, low redshift spirals. With a small jolt I realized that was very close to what I was finding now with the galaxy

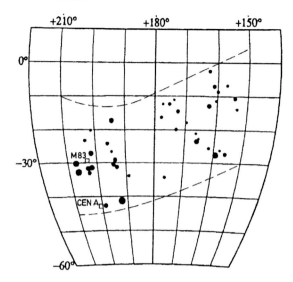

Fig. 6-16. A ROSAT survey
by Marguerite Pierre *et al.*
of X-ray emitting galaxy
clusters over the outlined
region in the southern sky.
The two brightest galaxies
in the region are marked.

clusters. Well, I had escaped ruin before, and the new results were nice support of the previous ones. I had started the letter of submission, when I thought about all the important results on ejected X-ray quasars that I was in the unique position to launch toward publication. What if my access to that data was shut off? At that point I put down my pencil and said, "Well, maybe it won't hurt to postpone submitting the galaxy cluster paper for a little while."

So here it is for the first time in this chapter. Later in the chapter I will indicate how the earlier publication on galaxy alignments supports these new cluster results.

A Complete Sample of X-ray Clusters

As soon as systematic X-ray observations of the sky began to be made it was discovered that many Abell Clusters were strong sources of hard, high-energy radiation. We will argue later that this marks them as young, active entities, but as a practical matter it enables X-ray observers to search systematically for galaxy clusters. As in the case of quasars, the X-ray surveys enable this class of galaxy clusters to be efficiently picked out of a large mass of background objects. In 1994 a complete survey of X-ray clusters over a large region of the Southern Hemisphere was published. *As Figure 6-16 shows, there is a striking concentration of X-ray clusters in the region, and the two brightest galaxies in this entire region are in the center of this concentration.* They turn out to be CenA and M83—*just the galaxies we found associated with the catalogued Abell Clusters in the beginning of this chapter* (Figure 6-7). This most recent survey then furnishes an independent and striking confirmation of those original results.

The most amazing thing about this investigation is perhaps the obvious non-random distribution of the X-ray clusters in this region of the sky and the failure of the investigators to comment on it. Perhaps the next most amazing aspect is that the largest grouping of the brightest X-ray clusters in this whole large region conspicuously coincided with the brightest galaxies in the region—but went unremarked. Why would

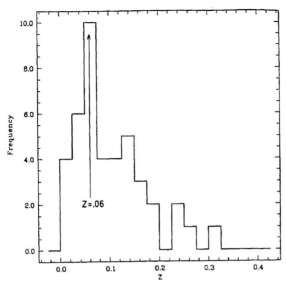

Fig. 6-17. The distribution of redshifts of the galaxy clusters in the previous figure shows a very strong peak at redshift z = .06. It is shown later that this is a characteristic redshift peak of quasars and AGN's.

the brightest X-ray clusters coincide with the bright galaxies in the CenA group if they were unrelated background objects? The remaining X-ray clusters in this area, though less bright, form two distinct strings; the ends of which also coincide with somewhat less bright, but still major galaxies (*e.g.* NGC3223 and NGC3521). One valuable aspect of this study, however, is that the investigators were obviously not previously biased in favor of finding galaxy clusters associated with nearby galaxies.

Another sensational result which was not noted was that the X-ray clusters in the surveyed area had an enormous redshift peak at z = .06. (Figure 6-17). The reason why this is such a key result is that the analysis of quasar and AGN redshifts showed the first quantization peak at about z = .061 (see following Chapter 8). Pencil beam surveys of field galaxies show this peak, but the active quasar-like objects show it much more strongly. *The presence of this peak with such great strength in the X-ray galaxy clusters marks them also as young, intrinsically redshifted objects.*

Blue Galaxies in High Redshift Clusters

A few clusters of galaxies with redshifts appreciably above the limit of the Abell clusters have been investigated in detail. Three with redshifts of z = .41, z = .42 and z = .54 have been described by R.J. Lavery, M.J. Pierce and R.D. McClure. They are outstandingly peculiar. They do not look anything like our galaxy or Virgo cluster galaxies. They are rich in blue and ultraviolet light, have increased fractions of emission-line galaxies and disturbed morphologies. In short, they are young galaxies—just the kind we have found empirically to have increasing amounts of intrinsic redshift.

When this Butcher-Oemler effect (as it is now called) was first discovered for clusters greater than about z = .2, it was a great shock to orthodox theory, which expected these "standard candles" to remain constant in their properties out to large redshifts. There was no alternative then for Big Bang adherents but to accept that at very short distances from us the universe changed drastically. This disaster was papered over by attributing it to "evolution." But no one faced the fact that the whole Big Bang solution of the general relativistic field equations rested on the assumption that the universe was homogeneous and isotropic.

For the empirical age-redshift law which we have been developing, however, the observed behavior of young clusters is what we would expect. The galaxies in them are low luminosity and high redshift when born from nearby, older galaxies. At a greater distance they are too faint to be seen. So these clusters are all nearby even if there existed more distant examples.

A clinching observation is that so many of these galaxy clusters are strong sources of hard X-rays. Galaxy clusters are supposed to be ensembles of old objects! How can they radiate so much energy for so long? The canonical explanation for this is "cooling flows"—big, old galaxies in the center of the cluster eject hot gas which emits X-rays as it cools and falls to the mass center. But this hypothesis is based on the assumption that the high-energy radiation is in equilibrium over the supposed 15 billion year age of the clusters—an assumption about which the most polite thing one can say is that it is unwarranted. In any case, after innumerable publications extolling cooling flows, it turns out they won't work.

"Cooling Flows" in Galaxy Clusters

A typical cooling flow has at least 100 solar masses per year. Over a billion years this would entail 100 billion solar masses. Where is it all hiding?

- One cannot hide this much star formation from optical observations.
- The gas cannot be warm molecular gas because it would be observed in millimeter wavelength emission lines.
- The gas cannot be cold molecular gas because the same wavelengths would reveal absorption lines.
- The last refuge is low-mass star formation, but such stars should show strong absorption bands in the near infrared.

Some cooling-flow theorists have admitted that the hypothesized mechanism leads to a catastrophe because it requires too much mass to be dumped on the central galaxy. They have tried to turn failure to fortune by arguing that the mass is thrown out again by jets and ejections. But this just raises a whole new set of problems of smoothness and lifetimes, and goes against the observations of clusters that have jets without any X-ray gas.

Clusters of Young Objects

The fact that these galaxy clusters characteristically emit high-energy X-rays marks them as young—just as the newly ejected quasars in the first few chapters are marked as young by, among other things, their rapid rate of X-ray production. And, of course, we have seen in the beginning of this chapter the X-ray cluster ejected from NGC5548 with z = .29 which is about the luminosity of an ejected quasar divided into 16 pieces.

The spectrum of an individual galaxy in one of these clusters usually reveals a composite of absorption lines from the stars it contains. As we have discussed in Chapter 4, these lines can tell us the stars are older than a billion or so years, but they cannot tell us whether they are as old as the 15 billion years required by the Big Bang. As calculated previously, a galaxy of matter created only two billion years later than our own would have an intrinsic redshift of about $z = .1$, but its oldest stars would only be 12% younger and this would not be readily detectable in a composite spectrum. As we have pointed out in this Chapter, however, when galaxy cluster redshifts approach $z = .2$ the constituent galaxies (shockingly) start turning very blue and showing non-equilibrium shapes—a sure sign of young galaxies.

There has never been a careful discussion or detailed sequence of comparisons of individual galaxies to check the assumption that faint groups of fuzzy smudges of light were the same as nearby great clusters, only seen at a greater distance. One could see a blatant difference when comparing the X-ray properties of fainter clusters to great clusters like the Virgo Cluster and A1367 (a continuation of the Virgo Cluster northward in the supergalactic plane), because in the latter clusters the individual large galaxies were X-ray sources. In the vast number of much smaller X-ray clusters, the whole group of objects was embedded in a diffuse X-ray emitting medium. The latter was a much different kind of cluster. Moreover, a cluster like the Coma Cluster did not even contain galaxies that looked like the giant Sb's and Sc's in Virgo. They were essentially just star piles called "E" (for elliptical) galaxies. The continuing assumption that faint little fuzzy spots in clusters of vastly smaller angular extent are the same kind of galaxies as populated the Local Supercluster seems not terribly good judgment in view of all the evidence for anomalous redshifts.

The Fornax Cluster Revisited—BL Lac Objects

In a final synthesis of what we have been talking about so far—quasars, BL Lac objects and clusters of galaxies—we return to the Fornax Cluster of galaxies. We take the same plot of bright Fornax galaxies and bright Abell clusters as shown previously, and now plot the positions of all known BL Lac objects in this large area of the sky (Figure 6-18). The first point of interest is that two of them (HP is a high polarization object in the BL Lac class) fall (with four Abell clusters) in the NE ejection cone from NGC1097. A new BL Lac object (bright and stronger in X-rays than NGC1097 itself as discovered from ROSAT survey scans by Arp and Fairall) then falls in the SW ejection cone. There seems to be a clear ejection of quasars, BL Lac's and galaxy clusters from this very bright, very active Seyfert galaxy with the striking optical jets.

The other two BL Lac objects in this field fall close to the most populated portion of the Fornax cluster. One of them we have discussed before in Chapter 2 as falling only 12 arc min from the spectacular barred spiral, NGC1365 (Figures 2-13a,b and c). The observers who measured the neutral hydrogen in this galaxy mentioned that it was extended in the direction of the center of the Fornax Cluster. If they meant NGC1291, that would be position angle = 212 deg. But at position angle = 203 deg. is the BL Lac

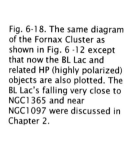

162   Clusters

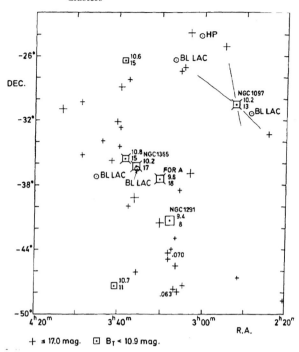

Fig. 6-18. The same diagram
of the Fornax Cluster as
shown in Fig. 6 -12 except
that now the BL Lac and
related HP (highly polarized)
objects are also plotted. The
BL Lac's falling very close to
NGC1365 and near
NGC1097 were discussed in
Chapter 2.

object which is much, much closer. It would be reasonable to conclude that the
hydrogen in NGC1365 had been entrained by the outward passage of the BL Lac
object. The fifth BL Lac of z = .165 falls close to the major concentration of galaxies
and Abell clusters in the Fornax Cluster.

It is clear again from Figure 6-18 that the Fornax Cluster represents a hierarchy of
redshifts. The largest galaxy is the oldest and has the lowest redshift. Successive
generations of galaxies are smaller, increasingly young and active and climb to higher
intrinsic redshifts in steps. The younger galaxies emerge in opposite ejection directions
and, with some rotation, give, as in Virgo, an overall "S" (for spiral) shape to the cluster.

The arrangement of known quasars within the ejection cone from NGC1097, and
the occurrence of the BL Lac objects and Abell clusters further out in these cones,
suggest the possibility of a very *important generalization* of the properties of ejected
objects. The suggestion is that the higher redshift quasars are the initially ejected objects,
and that as they travel outward their redshifts decay and their luminosities increase (as
their particle masses increase). By the time they reach a degree or so from their galaxy of
origin they have evolved into the bright apparent magnitude, z = .3 range where they
are liable to emit a burst of energy which transforms them into a short-lived BL Lac
phase. Perhaps at that time they can fission into a number of smaller parts and become
the Abell clusters which go on evolving to lower redshifts with increasing luminosities
for their individual galaxies. A schematic representation of this evolution is pictured in
Figure 9-3 in Chapter 9.

In any event the objects as they are plotted in Figure 6-18 establish at a glance that they are all physically related—and that the smaller more active objects have increasing amounts of intrinsic redshift. It is a picture where one has to either show that the plotted objects are not the brightest objects in their class, or bite the bullet and accept the observational result. *When looking at this picture no amount of advanced academic education can substitute for good judgment; in fact it would undoubtedly be an impediment.*

BL Lac Objects as the Progenitors of Galaxy Clusters

By now it is observationally obvious that there is an intimate connection between BL Lac objects and galaxy clusters. Since the BL Lac objects are at redshifts intermediate between quasars and cluster galaxies, in an evolutionary interpretation of redshift, the BL Lac objects would have to be the progenitors of clusters of galaxies. The key question is whether there are any observations which validate this inference.

The interesting thing about empirical observations is that they tell us that the BL Lac's break up (see the X-ray BL Lac at 1213 cts/ks above NGC5548 in Figure 2-3), and they tell us *how* they do it! Just as in the ubiquitous ejections that accompany the formation of young stars in our own galaxy (see Plate 8-19), the BL Lac's eject material in opposite directions. Apparently they eject a lot of it, and it eventually ages into smaller and somewhat higher-redshift companion galaxies and finally into clusters of similar redshift objects.

This result was foreordained when one considers that BL Lac's were shown by John Stocke and collaborators to occur in "cluster environments." Since the BL Lac's are associated with nearby galaxies (Chapter 2), the clusters must also be. For example the BL Lac-type quasar 3C275.1 (z = .557) was shown in Chapter 1 to be linked to a low redshift galaxy—Stocke also has shown that this quasar belongs to a cluster of galaxies. (See also Figures 8-11 through 8-13 for a cluster of quasars in the process of evolving into a cluster of galaxies).

The Twin Great Galaxy Clusters—Virgo and Fornax

The Fornax Cluster is not as well known because it is in the Southern Hemisphere and not studied as intensively as the Virgo Cluster. But for the redshifts which are known, it is fascinating to note their similarities with the patterns in Virgo.

As noted previously, in the Fornax Cluster the largest central galaxy and the strong radio galaxy, ForA, show the same pattern as in the Virgo Cluster with M49 and VirA. In Fornax two strong radio galaxies extend in a line away from ForA (as M86 and M84 extend away from Vir A). Again in the Fornax Cluster as in the Virgo Cluster the radio galaxies and spiral galaxies have systematically higher redshifts. It is particularly interesting to note that the luminosity class I spiral, NGC1365 (SBbI), is +824 km/sec with respect to the central Fornax galaxy (NGC1291). The four ScI spirals that are members of the Virgo Cluster are +582, +642, +824 and +1479 km/sec higher than the redshift of the brightest galaxy in the cluster, M49. (See also Chapter 3 and Figures 3-18 and 3-19.). So it is clear that the ScI spirals are a class of galaxies with well-established

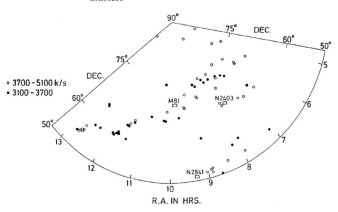

Fig. 6-19. All galaxies in the indicated redshift ranges are plotted over this enormous area of the sky. The brightest galaxy in the region is the well known M81, the next brightest NGC2403 and NGC2841 shown in the next figure.

intrinsic redshifts. There is also a concentration of quasars in the Fornax region (*Astrophysical Journal* 285, 555).

We have seen in both Fornax and Virgo how the Abell clusters outline the S-shape down the spine of the cluster. We have also seen how the BL Lac objects fall along this distribution in Fornax. It suggests that we look at the distribution of BL Lac objects in Virgo. When we do, we find two a little below and a few degrees to either side of M49. Their redshifts are .136 and .150, a good match for the .165 redshift BL Lac in Fornax. The two clusters are such a good match in all details, including the hierarchy of intrinsic redshifts, that I am tempted to say that if there is a creator (and if so I would not presume to attribute anthropomorphic properties to it) we might expect to hear: "*Look you dummies, I showed you the Virgo Cluster and you did not believe it so I will show you another one just like it and if you still don't believe it—well let's just forget the whole thing.*"

Galaxy Alignments

If anyone plots galaxies on the sky as a function of their redshifts it is clear that they form long, irregular strings. The most amazing property of these linear distributions is that the brightest, lowest redshift galaxy in the region falls in the middle of each of these strings. The situation is shown in Figure 6-19 for the longest string (next to the Perseus-Pisces filament) that I know of. There the string of 3100 to 5100 km/sec galaxies stretches more than 40 deg. across the northern sky. Right in its center is situated the giant Sb spiral M81, the next nearest major galaxy to our own Local Group, and having a redshift only a little over 100 km/sec.

Just to the west, M81 has a companion Sc, NGC2403, with its own shorter string of higher redshift galaxies. Then to the south we see two other strings of galaxies best outlined in the 4200 to 5200 km/sec redshift range. Figure 6-20 shows that each of these latter strings has a large, low redshift Sb at its center. In the Indian *J.Astrophysics Astron.* 11, 411, 1990, I was able to investigate the 20 brightest apparent magnitude spiral galaxies north of Dec. = 0 deg. Of the 14 which were uncrowded by nearby bright galaxies, a total of 13 had well marked lines and concentrations of fainter, higher-redshift galaxies. It is embarrassing to have to report that the chance of this occurring

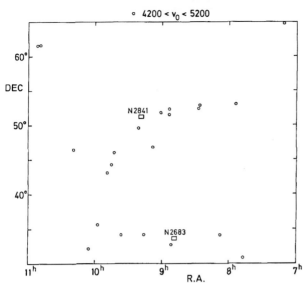

Fig. 6-20. All galaxies in the indicated redshift range plotted in the region south of M81. The two brightest apparent magnitude galaxies are indicated.

by accident is less than about one in 10 billion to 1000 billion. The calculation was made at the request of a referee who afterwards appeared still not willing to believe the plots of the catalogued galaxies.

Rather than repeat all the plots, I will show just one more in Figure 6-21. All the galaxies listed in the very complete *Revised Shapley-Ames Catalogue* by Sandage and Tammann are plotted here. There happens to be a scarcity in this particular region, so the bent alignment of galaxies is very conspicuous. Again, just at the middle of the string is situated the brightest, low redshift galaxy in the whole area (again an Sb).

Two of the galaxies in the string are ScI's, a particularly young kind of spiral which, it was shown previously, have particularly conspicuous excess redshifts. This brings up the important point that in the cases of the original 20 brightest spirals, investigation of the kinds of galaxies in the strings turned up an outstanding number of ScI's, disturbed, irregular, double, peculiar and active galaxies such as Markarian and Seyfert galaxies. This is clear additional confirmation of the non accidental nature of the strings, and it also points strongly to the conclusion that the redshifts of the aligned galaxies are intrinsic and caused by their younger age.

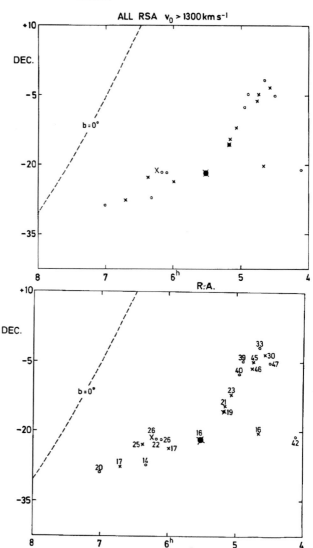

Fig. 6-21. All galaxies with redshifts greater than 1300 km/sec from the Revised Shapley-Ames Catalog (Sandage and Tammann) are plotted. The bottom panel appends their redshifts in hundreds of km/sec. The central, brightest galaxy is an Sb morphological type, crosses are later type spirals, large cross is an ScI type.

Alignments, Quasars, BL Lac's and Galaxy Clusters

Without going into the ritual models and detailed calculations of science, we can induce quite a bit of understanding by just noting the empirical relationships in the data we have discussed so far:

1) Objects which appear young are aligned on either side of eruptive objects. This implies ejection of protogalaxies.

2) The youngest objects appear to have the highest redshifts. This implies that intrinsic redshift decreases as the object ages.

3) As distance from the ejecting central object increases, the quasars increase in brightness and decrease in redshift. This implies that the ejected objects evolve as they travel outward.

4) At about z = .3 and about 400 kpc from the parent galaxy the quasars appear to become very bright in optical and X-ray luminosity. This implies there is a transition to BL Lac Objects.

5) Few BL Lac objects are observed implying this phase is short-lived.

6) Clusters of galaxies, many of which are strong X-ray sources, tend to appear at comparable distances to the BL Lac's from the parent galaxy. This suggests the clusters may be a result of the breaking up of a BL Lac.

7) Clusters of galaxies in the range z = .4 to .2 contain blue, active galaxies. It is implied that they continue to evolve to higher luminosity and lower redshift.

8) Abell clusters from z = .01 to .2 lie along ejection lines from galaxies like CenA. Presumably they are the evolved products of the ejections.

9) The strings of galaxies which are aligned through the brightest nearby spirals have redshifts z = .01 to .02. Presumably they are the last evolutionary stage of the ejected protogalaxies before they become slightly higher redshift companions of the original ejecting galaxies.

The connection between these dots of facts, which reveals the whole picture, seems to be that newly created, high redshift material is ejected in opposite directions from active galaxies. The material evolves into high redshift quasars and then into progressively lower redshift objects and finally into normal galaxies. This summary conclusion is elucidated further in the theory in Chapter 9, but here it is an example of a working hypothesis. It can be used to deduce what kind of physical processes are required to produce the observed effects—*i.e.* which theory out of the infinite number of possible theories is closest to reality. As a working hypothesis it stands ready to be amended any time better observations become available.

For example, it is not clear whether a large amount of newly created material will evolve in the same way as a small amount. It is possible that new material will be trapped in the interior of an originating galaxy, condense and then be ejected out in a later event in a more evolved state.

In my opinion, the above is almost exactly opposite the way current academic science works. Regardless of how scientists think they do it, they start with a theory—actually worse—a simplistic and counter-indicated assumption that extragalactic redshifts only mean velocity. Then they only accept observations which can be interpreted in terms of this assumption. This is why I feel it is so important to go as far as possible with empirical relations and conclusions. This is why it so important to discard any working hypothesis if it is contradicted by the observations—even if there is no alternative hypothesis to replace it. As unpleasant as it is, one must be able to live with uncertainty. Or, as many people say, but do not believe, "It is never possible to prove a theory, only to disprove it."

Chapter 7

GRAVITATIONAL LENSES

Prior to the 1950's Fritz Zwicky, the Swiss astronomer who had an illustrious and turbulent career in California, was aware that strong gravitational fields had been shown to bend light rays—as in the famous eclipse observations of the displacement of positions of stars observed at a grazing angle to the sun's limb. At that time he started looking for an extragalactic object which might be directly behind another, and thus have its outer light rays bent inward by the gravitational field of the foreground object so that it formed a ring or halo. Some "ring galaxies" were found, but they all seemed to be physical rings around the galaxy and not magnified background objects.

The more common situation to be expected was when the background object was not exactly centered and the gravitational ring collapsed into a one sided arc. But no striking examples of that were found either, so the subject had gone dormant. The sudden revival of gravitational lensing to the huge industry it is today is simply due to the quasars. In the 1960's and 70's I started finding high densities of quasars concentrated around nearby, low-redshift galaxies. Because of their high redshifts, it was felt that they could not be associated with low-redshift galaxies. As described in *Quasars, Redshifts and Controversies,* the observations were simply rejected as being incorrect. Then a theoretician named Claude Canizares got the idea that these apparent associations might be background quasars magnified in brightness by the gravitational lensing effect of the foreground galaxy. Suddenly the observations were hailed as important and correct, and many more examples of concentrations of quasars around lower redshift galaxies were found.

Excess Quasars around Galaxies

It was nice to be a hero but it was only for a day. That was all right with me though because I thought the pairing and separations of the quasars made their ejection origin from the galaxies indisputably obvious. It was on, then off, for galaxy lensing,

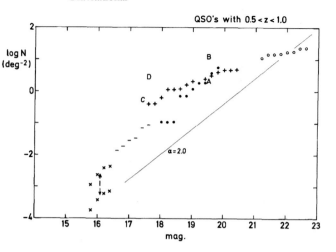

QSO's with 0.5 < z < 1.0

Fig. 7-1. Cumulative counts of quasars as a function of limiting apparent magnitude. The different symbols represent results from different surveys. Points B and D represent points from Arp showing excess quasars around nearby galaxies. The line at slope 2 represents the minimum rate of increase of quasar numbers required for gravitational microlensing.

then on again for microlensing (lensing by small objects such as stars and planets within the galaxy). When I heard that the gravitational microlensing calculations required a steep increase of quasar numbers with fainter apparent magnitudes, however, I protested that the observed numbers flattened off as they became fainter. Figure 7-1 shows that only the brightest quasars have an increase steep enough to satisfy the predictions. When I submitted this to the *Astronomy and Astrophysics* Journal, the Editor was not about to believe it and it required a letter from one of the leading lens theorists, Peter Schneider, to convince him that the argument was correct. I will always admire Peter for his integrity in writing to support an opposing result.

But what would I predict for the number counts for quasars as a function of their apparent magnitude? That was actually a pretty easy question since if the quasars belonged to bright nearby galaxies, they would be distributed in space the same way. Figure 7-2 shows how the numbers of luminous Sb spirals like M31 and M81 increase with apparent magnitude (the crosses). The line segments show how the various quasar surveys increase with apparent magnitude. *The fit is extraordinarily good*, especially

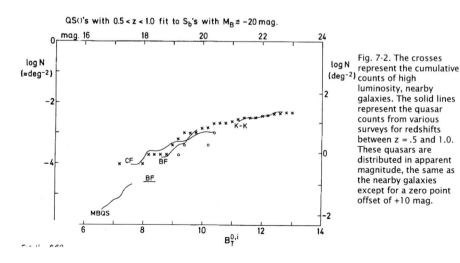

QSO's with 0.5 < z < 1.0 fit to Sb's with $M_B \lesssim -20$ mag.

Fig. 7-2. The crosses represent the cumulative counts of high luminosity, nearby galaxies. The solid lines represent the quasar counts from various surveys for redshifts between z = .5 and 1.0. These quasars are distributed in apparent magnitude, the same as the nearby galaxies except for a zero point offset of +10 mag.

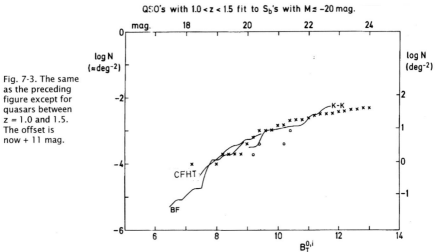

Fig. 7-3. The same as the preceding figure except for quasars between z = 1.0 and 1.5. The offset is now + 11 mag.

considering the non-linear shape of the two functions. Even the details fit well.

The adjustment in apparent magnitude required to make this fit is very significant. For quasars $0.5 < z < 1.0$, the displacement is 10 magnitudes. That means that quasars of the order of 19th apparent magnitude must be found around dominant galaxies of about 9th apparent magnitude. *This is exactly what had been observed!*

In Figure 7-3 the fit for quasars of $1.0 < z < 1.5$ is shown to be also very good but now at a zero point shift of 11 magnitudes. This supports the general finding that quasars become less luminous as they approach redshifts of z near 2.0. It would imply that if the number counts of z near 2.0 quasars were better defined, they would require about a 13 magnitude difference with their associated galaxies. That would make 18th to 20th magnitude, z = 2 quasars generally visible in association with only the nearest galaxies like M31, the Sculptor Group and M81. This also is what had been found in the earlier investigations as summarized in *Quasars, Redshifts and Controversies*.

My paper detailing the above analysis (*Astronomy and Astrophysics*. 229, 93, 1990) lists five independent reasons why gravitational lensing cannot account for the excess number of quasars around bright galaxies. But most decisively, it demonstrates that the observed number counts for quasars can *only be accounted for by their physical association with bright nearby galaxies*. This would reasonably seem to have settled the question. But in a paper in *Astronomy Journal* 107, 451, 1994 two authors reported a statistical association of quasars with "foreground" galaxy clusters—one of numerous recent papers reporting quasar/galaxy associations. They make the curious statement: "We interpret this observation as being due to statistical gravitational lensing of background QSO's by galaxy clusters. However, this ... overdensity ... cannot be accounted for in any cluster lensing model ... and is implausible in any conventional model of cosmic mass distribution." Most startling, they do not even reference the *A&A* paper from which we have excerpted here the figures which empirically disprove gravitational lensing for quasars. As papers multiply exponentially one wonders whether the end of communication is near.

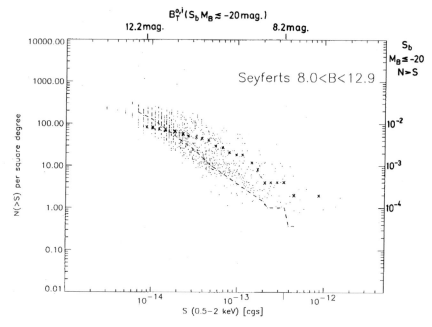

Fig. 7-4. The cumulative numbers of X-ray sources around Seyferts as in
Fig. 2-1. The high luminosity Sb's are plotted as crosses showing again
the same distribution in apparent magnitude as the excess
(predominantly quasar) X-ray sources.

Distribution of Quasars around Seyferts

We can test the distribution of X-ray quasars around Seyfert galaxies which we
found in Chapter 2 by means of the same number count diagram used above. Figure 7-4
shows the cumulative counts for the brightest Sb galaxies again as crosses. It is seen that
they follow very well the excess numbers of X-ray sources associated with the Seyferts.
This is as it should be because the bright Sb's are basically the progenitors of the more
active Seyferts and are distributed in space the same way. Likewise the quasars that are
associated with these galaxies show the same number count behavior with apparent
brightness.

In fact these number count diagrams become the only statistical method of
measuring extragalactic distances if the redshifts are not velocities. Therefore they
should become very important tools as time goes on. For example at the moment no
one has the faintest idea where the mysterious gamma ray bursters are located. But their
number counts have a characteristic break resembling the break in the number counts
for $z = 1$ objects. That means they are probably distributed in space like the active $z = 1$
quasars, that is throughout the Local Supercluster.*

---

*   As this book goes to press the gamma ray bursters appear to be associated with faint, active galaxies, one
      about $z = .8$.

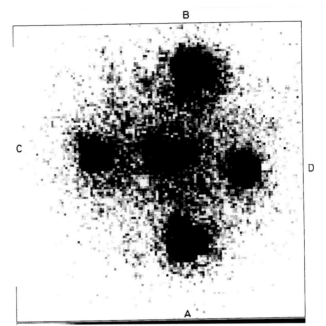

Fig. 7-5. The Einstein Cross—four quasars of redshift z = 1.70 aligned approximately across a central galaxy of redshift z = .04. First publicly released picture from the Hubble Space Telescope of this object.

The Einstein Cross

The most celebrated case of a galaxy supposedly splitting the image of a background quasar into separate images by means of its gravitational field is a rather unimposing object called G2237+0305. When it was first discovered it caused a panic because it was essentially a high redshift quasar in the nucleus of a low redshift galaxy (*Quasars, Redshifts and Controversies* p.146). Gravitational galaxy lensing *had* to be invoked for this one.

Subsequent high resolution observations showed four quasars of z = 1.70 in the form of a cross approximately centered on a 14th magnitude galaxy of z = .04. Since the four quasar images were all within one arc sec of the galaxy nucleus it was impossible to claim accident. But the gravitational lens was in big trouble from the start because Fred Hoyle quickly computed that the probability of such a lensing event was less than two chances in a million!

But by this time I had already been curious enough to look hard at the publicity picture which NASA had released of the space telescope picture of the "Einstein Cross" (Figure 7-5). It is an exemplary demonstration of how science proceeds when one realizes that thousands of scientists must have "looked" at that picture and said, "Oh here is quasar image that has been split into four parts by the action of one of those gravitational lenses." I was skeptical of gravitational lenses and I looked at the picture and saw what should not be there, a luminous connection between one of the quasars and the nucleus of the galaxy!

It is good to have friends you can trust to be honest. I took the picture to Phil Crane and said, " Do you see what I see?" After studying it for a little while he said,

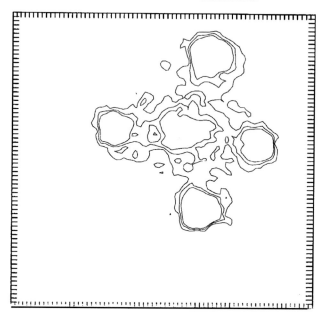

Fig. 7-6. The slightly smoothed, contoured image of the preceding picture showing the elongation and connection of the central galaxy to the east-west quasars and the extensions of the north-south quasars toward the galaxy.

"Yes, but I don't know if it's real." "Well, how do we find out?" I asked. Fortunately he was a member of the instrument definition team for space telescope imaging. He had all the exposure frames and processing tools in his computer. We contoured the image and tested against artificial stars superposed near the galaxy. Not only the original connection, but extensions toward the other quasars looked real! (Figure 7-6). *One should carefully consider the following important question: What is the chance that a person who notices an important discrepancy in a scientific announcement has the opportunity to check it out at the level of the primary data?*

In the course of processing the images, however, we realized that by a stroke of good luck there were three ultraviolet exposures with the space telescope in the wavelength band centered at 3400 Å. That band includes the Lyman alpha line of the redshifted quasars. This was the strongest emission line of the most abundant element in the quasars, and would be most likely to show any gaseous connection. Color Plate 7-7 shows the breathtaking result: the western quasar (D) is connected directly into the elongated galaxy nucleus! There is absolutely no way to escape the overall result that the quasars are connected and generally elongated toward the low redshift nucleus.

As if this was not enough, shortly afterwards I was walking back from an IAU session in Buenos Aires with Howard Yee at my elbow.

"What's new, Howard" I asked.

There was a long silence and then, "Well there is something you would probably be interested in", he murmured.

"What is it?" I politely asked.

"Well", after some hesitation he went on, "We put the slit of the spectrograph between quasars A and B in the Einstein Cross and we registered a broad Lyman alpha emission in each quasar. But between them we found a narrow Lyman alpha line—it looks like there is some low density gas at the same redshift as the quasars between them."

A jolt ran through me and I looked at him to try to read the expression on his face. As usual in such situations, his eyes avoided mine. The point was, of course, that a line between quasar A and B passed directly between the nucleus of the galaxy and quasar D. On the face of it *high redshift gas was indicated near the nucleus of the low redshift galaxy*. But what I knew, and what anyone can know looking at the Lyman alpha centered photograph in Color Plate 7-7, is that there is a putative Lyman alpha filament connecting quasar D to the galaxy nucleus. What the spectrum had confirmed was that this indeed was a low density, excited hydrogen filament connecting the two objects of vastly different redshift. We are again seeing trails of material resulting from ejection and, as we saw in the first few chapters, tendencies for orthogonal ejection from the parent galaxy.

Censorship at the Critical Point

Phil and I wrote all these results up in the best journalese style with all the tests, numbers and references and submitted it to the *Astrophysical Journal Letters*. The editor sent it to a referee who had just written a long paper on the Einstein Cross—he had been looking for the fifth image predicted by gravitational lens theory, had not found any convincing evidence for it, but concluded anyway that he had strengthened the interpretation as a lens. He wrote three reports trying to get us to say our results were all the result of chance noise and then rejected the paper. When I pointed out to the editor the conflict of interest with the referee's recent paper, the editor sent it to a theoretician with an even stronger conflict of interest. The latter essentially said that since it disagreed with current theory the observations must be wrong. The paper was finally published in *Phys. Lett. A* 168, 6, 1992.

I feel very strongly about what happened and I want to make my position clear: *Astrophysical Journal Letters* is the normal journal for publishing new observations from the HubbleSpace Telescope. The telescope cost billions of dollars of public funds. The vast majority of page charges which pay for the publication of the journal come from government supported contracts. The overriding, first directive of the editor is to communicate important new astronomical results. If the editorial process violates its primary responsibility, it misuses public funds.

More Scandals of the Cross

Papers which claim that the data confirms the standard theories, however, are rapidly published. One example of such a paper is Howard Yee's (1988) analysis which computed a required mass of about 100 billion suns inside the very small radius where the quasars are located. This leads to a mass to light ratio (M/L) of about 13 which Yee states "...is near the high end of that of large spiral galaxies... but is entirely acceptable." Well not quite. If you consult the original reference (Kormendy 1988), you see that this

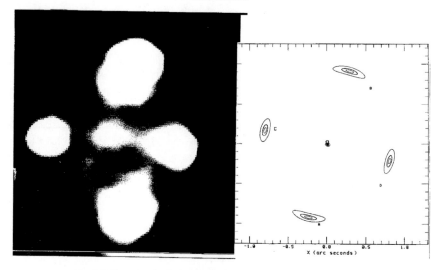

Fig. 7-7. Theoretical calculations by Peter Schneider *et al.* of what gravitationally lensed quasars should look like. If resolved the luminous isophotes should be extended by a factor of 4 or 5 to one along a circumference. Left panel shows HST observations.

M/L is completely above the bulges for spiral galaxies, even above that for E galaxies and that is for a *lower* limit on the required M/L. In fact if you extrapolate the luminosity required for an elliptical to have this M/L ratio it comes out M_B = -25 mag. How bright is -25 mag.? Well conventional quasars start at M_B = -23 mag. so this galaxy would have to be 2 magnitudes brighter than the supposed brightest objects in the universe, a conventional quasar!

In a later paper, (Rix, Schneider and Bahcall (1992) in a computation essentially no different from Yee's derive the same mass for the galaxy. Their contribution was to change "entirely acceptable" to "lens models .....explain the wealth of observations elegantly." Actually this so-called confirmation rests on the assumptions that the redshift dispersions in galaxy interiors represent both velocities and velocities in equilibrium. As we have seen the observations show this is incorrect on both counts and systematically overestimates the masses of galaxies. Furthermore the mass needed for lensing is underestimated by assuming it is all concentrated at a point in the center of the galaxy—hence the lower limit mentioned above. Also in the matter of the Cross a disk galaxy is called an elliptical and mass-to-light ratio comparisons are shuffled between blue, visual and red. Even after all this one requires an extraordinary and unprecedented galaxy to satisfy the lens requirements.

For realistic galaxy masses, gravitational lens effects may someday show up, but on a much smaller scale than currently claimed. As for the central galaxy in the Einstein Cross, one only needs to look at it to realize that it is in fact a small, dwarf galaxy. I think it would be an enormously helpful reality check if astronomers studied galaxies in groups and learned to judge the giant, medium, dwarf characteristics of a galaxy from its morphological appearance.

The General Case Against Lensed Quasars

The Arp, Crane paper on the Einstein Cross also made clear that a basic requirement of gravitational lensing had been violated by all the observations of quasars. The theoretical calculations in Figure 7-7 show that the lensed images would need to be elongated by a factor of 4 or 5 to 1 along the circumferential direction. This is just common sense, because as was mentioned in the beginning of this chapter, people originally started looking for gravitational rings around deflecting galaxies. In the more common case where the lensed object is not exactly behind the lens, the ring degenerates into segments of rings or arcs. The fact that the supposedly lensed quasars never looked like arcs was always excused on the basis that they were unresolved point sources, *i.e.* if one could get enough resolution then they would be seen as arcs. But Figures 7-5, 6 and Color Plate 7-7 of the Einstein cross show that the images are well resolved and, *instead of being arcs, the quasar images are extended back toward the central galaxy.*

As a result, lensing of any quasar was excluded from the beginning, because the theorists forgot their own theory that quasars were nuclei of host galaxies. Even if the nuclei could not be resolved, the host galaxy was supposed to be of the order of 40 kpc in diameter, and since its surface brightness was preserved under magnification, it should have been quite visible and noticeably arc shaped. Recall that 3C48, the first quasar discovered, was supposed to have a host galaxy of 12 arcsec in extent around it. 3C48 was about 16th apparent magnitude. Should we believe all the quasars of comparable redshift at 19th apparent magnitude with less conspicuous nuclei would not have had conspicuous host galaxies on the conventional assumption of redshift distance? And not an arc among them!

We are just about to get into the subject of high redshift arcs in clusters of galaxies. They are supposed to be hugely distant background galaxies lensed by the gravitational field of the cluster and drawn out into thin, conspicuous arcs. If this is true, the enthusiasts cannot have lensed objects at the same redshift appear both as point sources *and* arcs. They simply cannot have it both ways!

Arcs and Arclets in Clusters of Galaxies

When Roger Lynds first registered arcs in a deep photograph of a cluster of galaxies he did not pay much attention to them. Later they were highlighted by Vahe Petrosian but it was not until the redshifts of some of them were measured as being very high that the idea that they had to be gravitationally lensed arcs of distant background galaxies became mandatory. I was suspicious of this interpretation because I felt high redshift objects were generally not so distant and also because I felt the galaxy clusters had much smaller masses than conventionally estimated from the redshift dispersions of their members. But I had to admit that the fairly accurate arc shapes centered on the clusters looked like the expected degenerate Einstein rings and made a very persuasive case for gravitational lensing.

When the evidence for very small, nearby Abell galaxy clusters (which was presented in the preceding chapter) began to sink in—and it took some effort to change

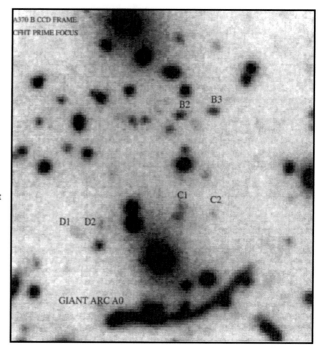

Fig. 7-8. A deep CCD image of the galaxy cluster Abell 370. A gravitational arc is marked as well as images supposed to be multiple images of background objects. Following three pictures adapted from B. Fort and Y. Mellier.

my perception of them—I realized it was impossible for them to have enough mass to lens any distant objects. Taking a more critical look at the observed properties of the supposed lensing clusters revealed some very surprising properties. For example, A370, pictured here in Figure 7-8 has a redshift of $z = .375**$. Not only is this close to $z = .30$ redshift peak for intrinsically redshifted quasars but if we were to try to fit it to the redshift-apparent magnitude diagram for Abell clusters, we would find it had an excess redshift of about 30,000 km/sec! Peculiar velocities of this order of magnitude are simply not contemplated for the conventional Hubble flow. Moreover A370 is only 1.7 deg away from the large, active Seyfert NGC1068 which was found to be a source of ejected quasar candidates in Chapter 2. In fact the elongation of the cluster points right back to NGC1068!

Another cluster with supposed gravitational arcs, A2281, has a more reasonable redshift of $z = .176$ for the apparent magnitude of its members. But that is the same redshift of some of the clusters and BL Lac objects running down the spine of the Virgo and Fornax Clusters. In fact A2281 is only about 1.3 deg away from a 14.46 mag. Seyfert galaxy of $z = .026$ and only 25 arc min away from a 14.3 mag. spiral of unknown spectrum.

Still another cluster with supposed gravitational arcs is MS0440+02, pictured in Figure 7-9. The arcs don't look like an elongated background galaxy, they look like an ejected shell. What is more, in this case there is a Seyfert galaxy only 22 arc min away

---

** B. Fort and Y. Mellier actually say about these arc clusters "Note the strong peak between $z = .2$ and $.4$." We will show in the following Chapter 8 that these clusters have the same quantized redshift distribution as BL Lac's and quasars.

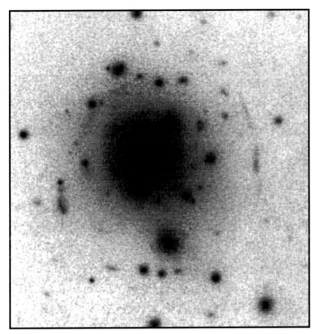

Fig. 7-9. The X-ray cluster MS0440+02 showing supposed gravitational arc to the west of the large central galaxy.

with essentially the same redshift (z = .196 vs z = .19). Obviously they are a pair of active galaxies but one has more companions plus a shell.

Just casual inspection shows these active clusters lie strikingly closer on the sky to low redshift and active galaxies than they should be by chance. Again it would be of the highest priority to systematically test the correlation between these clusters and catalogues of various kinds of galaxies and quasars.

The development that really shocked me, however, was the report of *radial* gravitational arcs—*i.e.* "arcs" that pointed back to the center of the cluster rather than tangential to it. That was exactly opposite to the expectation of degenerate Einstein rings!

Radial "Gravitational Arcs"?

Figure 7-10 shows the prize exhibit—the cluster MS2137-23 with a supposed "radial arc" emerging from the central galaxy and aimed directly at the middle of a "tangential arc." While I was trying to get over this latest example of "our theory thrives on adversity" bravado, something was nagging at my memory. Where had I seen this before? Then it struck me—the longest, straightest jets emanating from a spiral galaxy—NGC1097! Color Plate 2-7 shows in true color that the strongest of the four jets ejected from this extremely active Seyfert ends in a right angle. Dare I say tangential arc? The right angle turn of the main jet in NGC1097 was always a perplexing mystery. It was jokingly called the "dog leg jet." But it was obviously ejected—could it have hit a cloud or piled up material ahead of it and then flattened at a right angle to the direction of propagation?

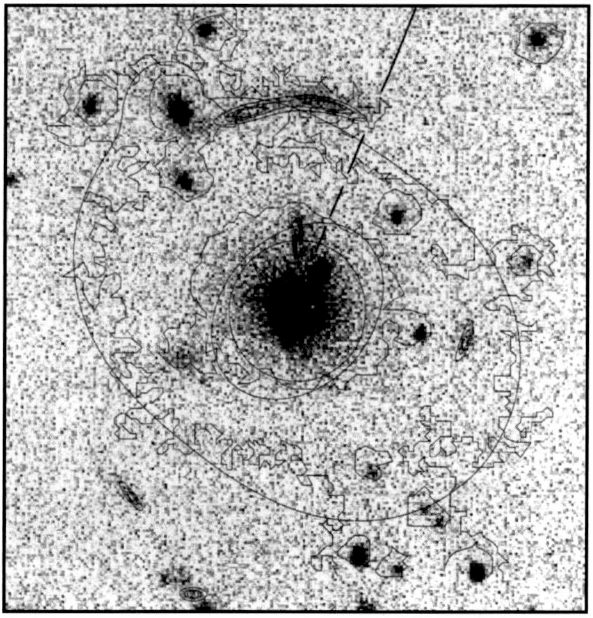

Fig. 7-10. Another X-ray
galaxy cluster, MS2137-23,
with claimed *radial*
gravitational arc pointing
directly toward a transverse
arc. A quasar of z = .646 lies
49 arc min away in direction
of the "radial arc."

Was there any other example like this? Why, yes, there was the second strongest jet from a spiral galaxy, the one from NGC4651. When I was reducing the X-ray observations shown in Chapter 1, I wanted to get the deepest possible photograph of the galaxy in order to understand the relation of the jet to the nearby, apparently ejected quasar. The best possibility was the deep (IIIa-J) survey in progress with the Palomar Schmidt telescope. I am extremely grateful to Bob Brucato for sending me a copy of this object before the Survey was formally released. The limiting exposure is shown here in Figure 7-11. Lo and behold! The ejected jet ends again in a tangential arc, just as in the NGC1097 case!

So one would conclude, empirically, that the galaxy cluster in Figure 7-10 was also ejecting a jet which was responsible for an arc of similar material at right angles to its end. Note also that this galaxy cluster is of the type where one galaxy is dominant and the remainder are very much companions to it—the kind often involved in ejection activity. Also note how the low surface brightness of the arcs in clusters render them very difficult to discover even with the deepest photography. Then match that with the characteristics of the ejected jets from the spiral galaxies just discussed which are also only discovered on the deepest photographs.

If you look for what is being ejected in Figure 7-10, it turns out to be a quasar of z = .646 (the cluster is z = .313). This quasar is at p.a. = -22 deg and only 49 arc min distant. In fact it comes out right along the fat jet (if that is what it is) which is visible in Figure 7-10. Notice that the central galaxy in this cluster is dominant. What you actually have here is an active galaxy, surrounded by many companions, which is currently ejecting in the direction of the arcs and the quasar! We shall see in the next chapter that

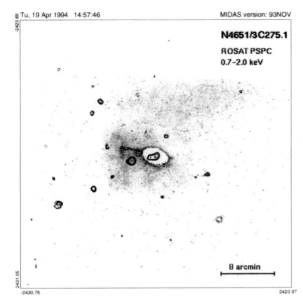

Fig. 7-11. A deep, IIIa-J exposure of the jet galaxy NGC4651 discussed in Chap. 1 (contours of X-ray sources superposed). Note the optical jet ends in a transverse, arc-like feature.

the redshift of the central galaxy in the cluster and the quasar are very close to the quantized redshift values of z = .30 and .60.

Diametric Arcs in a Galaxy Cluster

In a redshift measurement program of X-ray clusters discovered by ROSAT, the most luminous cluster was, on the conventional interpretation of redshift, RXJ1347.5 - 1145. (What this means in operational terms is that this cluster represents the most

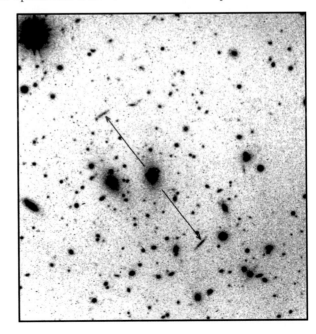

Fig. 7-12. The conventionally most luminous X-ray cluster in the ROSAT survey has short, diametric features. About 5 arc min NE of this cluster is a strong radio quasar of redshift z = .34. Photograph courtesy Sabine Schindler.

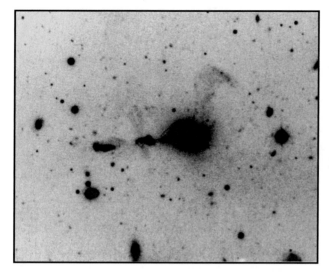

Fig. 7-13. Disturbed galaxy in the Herculis Cluster with a jet to the E which contains small galaxies. Note particularly arc to NNW with material from galaxy leading to it as well as fainter arcs to SW and W.

extreme deviation from the redshift—X-ray intensity, Hubble relation.) Possibly connected with this extreme property was the unique finding of two short arcs diametrically spaced across the more compact of the two bright central galaxies. The picture of this galaxy cluster and its arcs is shown in Figure 7-12. The eye is immediately struck by how short the arcs are and how exactly they are aligned across the galaxy. This is strongly similar to the quasar pairs ejected from active galaxies as presented in Chapters 1 and 2 and suggests the question: "What would happen if a quasar in formation or entrained material in the ejection encountered resistance in the direction of its motion? Would it flatten as it became more luminous?"

If it is a lens why are the arcs so short? And why is the circular symmetry broken in this opposed, even fashion? What about the equal spacing of the two across the galaxy?

By some (unexpected?) coincidence there is a bright (V = 18 mag) quasar (z = .34) about 45 arcmin NE of the compact central galaxy. The alignment is within about 15 degrees of the alignment of the two short arcs. Since the redshift of the cluster is z = .451, the implication would be that it was ejected from the quasar (as a strong radio source the quasar may have BL Lac properties). Regardless of the significance of this latter association, however, the cluster is a prototype of the ones we were discussing in the previous chapter. It is extremely strong in X-rays, located in the supergalactic plane between Virgo and Centaurus (see Figure 6-16) and just the kind of object we would consider to be a young, active object ejecting and breaking up into smaller, intrinsically redshifted young galaxies.

Galaxies which Eject Shells and Arcs

At this point we would like to show some pictures of real galaxies which are clearly ejecting material which uncannily resembles the claimed gravitational arcs.

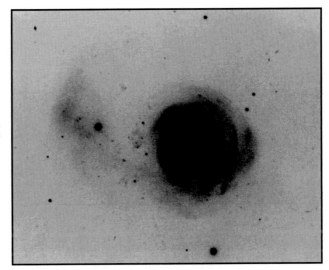

Fig. 7-14. A galaxy from *The Atlas of Peculiar Galaxies* (No. 215) which shows three distinct arcs on one side and an ejected plume on the other side.

First there is an extended cluster of galaxies called the Hercules cluster, which contains many disturbed and active galaxies. Figure 7-13 shows one of these which has a conspicuous jet emerging from it. There is no question that this feature belongs to the galaxy because the knots in the jet have more or less the same redshift as the galaxy. Viktor Ambarzumian in the 1950's presented this as an example of new galaxies being formed in an ejection. But of key importance for the present demonstration is the thin arc of luminous material to the NNW of the galaxy. This arc is actually connected back to the galaxy by a diffuse filament of luminous material. It is clearly also an ejected feature. Closer to the galaxy on the SW side is a fainter, thinner arc which continues around very faintly in a roughly circular fashion as if it might join the stronger arc to the NW. There are indications that deeper photographs would reveal further luminous arcs which are essentially identical to the proposed gravitational arcs in clusters but in this case are clearly a result of explosive activity in the central galaxy.

Figure 7-14 shows another galaxy that has three distinct arcs on one side and a long, presumed ejection tail in the opposite direction. The three arcs are evenly spaced and concentric, suggesting a ripple or vibrational impulse from the center of the galaxy.

As a final example, Figure 7-15 shows a deep photograph of a galaxy which has such sharp edged, circular arcs that it suggests a bell ringing or vibrating under water. Similar galaxies, and particularly this one, have been modeled as two galaxies merging. But in an encounter, the best one might expect is one galaxy spiraling into another. As noted, however, the pictured arcs are not spiral but perfectly circular. Moreover, as the picture shows, there are at least two faint jets emerging from the center which testify directly to the ejecting explosive nature of the nucleus.

Ejection of Arcs from Seyferts

The remaining question is whether ejected arcs can be composed of material which is of considerably higher redshift than the ejecting galaxy. One would naturally

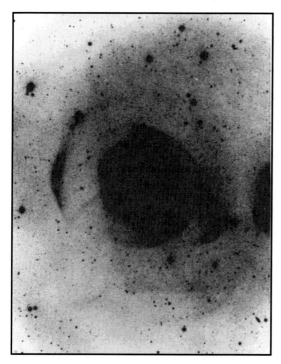

Fig. 7-15. Another galaxy from the *Atlas* (No. 227) which shows sharply defined arcs plus two faint jets coming out of the center.

think of the Seyfert galaxies which eject quasars of the same order of redshift as the supposed gravitational arcs measured in the galaxy clusters. Two investigators, Robert Fosbury and Andrew Wilson, among others have shown that Seyfert galaxies characteristically eject material in opposite directions in cones of various apparent opening angles. The most recent and spectacular example of this is shown in Plate 7-15.

The crucial aspect of this observation is that it shows that the material in the ejection cones is coming out in a series of concentric arcs. The arcs appear to be getting thinner as they progress outward, perhaps as a result of shocks or compression by faster following material. When a high redshift quasar or proto quasar is being ejected there may well be similar age material which trails the quasar and forms arcs. Or the quasar itself with low particle masses may be deformable into an arc if it meets clouds or medium in the neighborhood of the galaxy. (The last would be a possible model for the previously discussed short pair of arcs across the galaxy in Figure 7-12.)

Of course some arcs would be expected to be entrained material of the ejecting galaxy and some arcs new, young material of high redshift. So it would require a careful spectroscopic observational program on these kinds of objects in order to check this working hypothesis and modify it if necessary. The important result, taking everything so far, seems to be that active galaxies eject high redshift quasars and also eject diffuse material, some of which is in the form of arcs. Since many Abell clusters of galaxies are energetic X-ray sources and contain active galaxies, or galaxies perhaps recently active, it is reasonable to suppose they also could eject material that would appear as high redshift arcs.

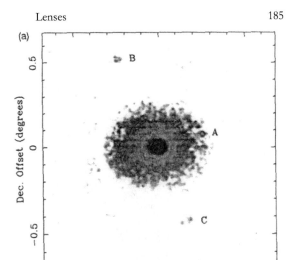

(a)

Fig. 7-16. An X-ray map of the Centaurus Cluster of galaxies showing the bright radio galaxy at the center, NGC4696, at z = .00975. The fainter companions average +375 km/sec higher redshift. The paired X-ray sources B and C await optical spectra. ROSAT map by S.W. Allen and A.C. Fabian.

The small size of the clusters which we found in the preceding Chapter 6, however, would seem to preclude the ejection of anything initially massive. Regardless of which is the optimum new model, however, the observations clearly confound our conventional assumptions.

Ejections from Abell Clusters of Galaxies

Because I was alerted by all the pairs of X-ray quasars we had been finding across Seyfert galaxies; when I first ran across the X-ray map of the cluster of galaxies shown in Figure 7-16, all I could see was the pair B and C across the central, active galaxy. (Actually, this is one of the best known galaxy clusters, the Centaurus Cluster). Since the two sources are aligned within the accuracy of measurement across the central galaxy, it is highly probable that they have been ejected.

I immediately took the listed positions of these sources and processed them in the automatic plate measuring program. Source A is identified with an 18.7 mag. BSO, possibly variable. Source B is identified with a 15.6 mag. blue compact galaxy almost certainly of much higher redshift than the central galaxy. Source C is either a 16.7 mag galaxy or a 17.1 mag BSO (fuzzy?) or a 20th mag object. It will require a spectroscopic program of observation to obtain redshifts and identifications and perhaps a check position on C. Such observations could easily be obtained with current 10 meter class telescopes, but this is the last thing one could expect to happen because they are completely occupied with really important projects. I may go to South Africa to try to measure the candidates with a 1.9 meter telescope. It is the same telescope I used to set distance scales to the Magellanic Clouds 40 years ago—moved to new location, with an aluminized mirror instead of silvered and a modern spectrograph. It would be some satisfaction to leapfrog the intervening years to a better understanding with the same telescope.

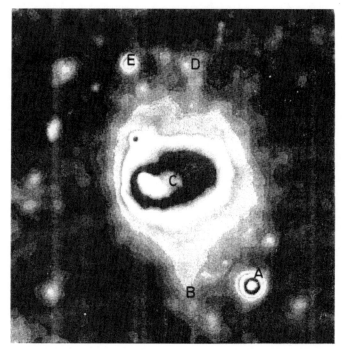

Fig. 7-17. The galaxy cluster Abell 754 in X-rays. The cluster has a redshift of z = .0528. The source E has z = .253 and D, attached to the cluster by a filament, has z = .129. Opposite E is A which is identified with a moderately bright galaxy. Opposite D, on the end of the pointed extension from the cluster is B, identified with a BSO. Rosat map by J.P. Henry and U.G. Briel.

There are several important comments that should be made about this cluster. The first is that it is a cluster of galaxies dominated by one active radio galaxy, NGC4696. The redshifts of the rest of the galaxies in the cluster average +375 km/sec higher. This is the same situation we found for all groups involving companion galaxies in Chapter 3. Secondly the cluster is the rich, bright Centaurus Cluster which was specifically associated with the giant CenA galaxy in Chapter 6. One can see the cluster in Figure 6-7 labeled z = .011, almost due west of CenA. As Figure 7-16 shows, the Centaurus Cluster is elongated back in the direction of CenA in the same direction as one of the two exceptionally strong, opposing absorption lanes which emerge from the center of that extremely active galaxy. The implication, as in previous cases, is that CenA ejected NGC4696 along the line of that lane and NGC4696 then gave rise to the surrounding cluster which is in the nature of later generation companions to it. It is perhaps significant to note that there is a 13.9 mag Seyfert galaxy of z = .016 near NGC4696, on this same line back to the western lane in CenA.

Abell Cluster A754

This cluster is so spectacular an X-ray object that it was featured in the ROSAT Calendar for 1995. It was there that the pairings of X-ray sources across it leaped out at me. Figure 7-17 shows that the bright sources A and E are paired across the center of the X-ray emission. E is catalogued as an 18.0 mag z = .253 active galaxy. (The redshift of the cluster is z = .0528.) A can be identified with a bright galaxy. It will be fascinating to obtain the redshift of that galaxy since it is obviously the counter ejection object to E.

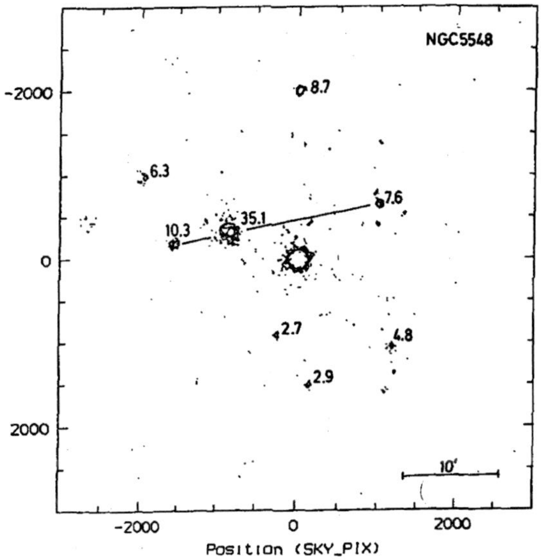

Fig. 7-18. An enlargement from the X-ray map around the Seyfert galaxy NGC5548. The galaxy cluster with the X-ray emission of 35.1 counts/kilosec has z = .29 and is shown ejecting a pair of quasars of z = .67 and .56.

But even more sensational is the X-ray source D. *It is connected back to the cluster interior by a luminous X-ray filament!* It is catalogued as an 18.3 mag active galaxy with z = .129. Naturally I looked for the counter ejection and there it was—an extension outlined by many X-ray contours leading to a sharp point at about B in Figure 7-17. On the deep Schmidt photographs I looked in that vicinity for a candidate object. Just as I was about to give up, at a little further along the line than I was looking, I spotted a double star one component of which was blue. I am betting that is the quasar opposite D. Of course, it has to be confirmed.

So now we have a cluster from which *two* ejections of high redshift objects have taken place. This and the previous case of an ejecting cluster were found serendipitously. It obviously would be highly profitable to systematically examine strong X-ray clusters for similar cases. An example of one such possible finding is the following:

The Galaxy Cluster Associated with NGC5548

In the systematic survey of bright Seyfert galaxies discussed in Chapter 2 one very strong X-ray cluster was found apparently ejected from the very strong Seyfert NGC5548. Figure 7-18 here shows the archived X-ray map processed by Arp and Radecke. It is clear that a strong pair of X-ray sources is paired across this galaxy cluster of z = .29. The quasars happen to be catalogued and have redshifts of z = .67 and .56. As discussed previously this correspondence in properties essentially assure them of being an ejected pair from the galaxy cluster. (Of course there may have been a single object at z = .29 at the time of the ejection of the z = .67 and .56 pair. In the act of ejection it may have broken into the small cluster objects we presently see).

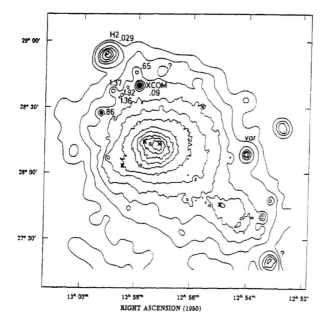

Fig. 7-19. An X-ray map of the Coma Cluster of galaxies, the most conspicuous cluster known of predominantly non-spiral galaxies. X Comae is a z = .09 Seyfert 1 galaxy much like those found in the Virgo Cluster. Point X-ray sources identified as quasars are marked with their redshifts. X-ray map from S.D.M. White, U.G. Briel and J.P. Henry.

Although many more cases need to be investigated, the results so far seem to assure the phenomenon of galaxy clusters ejecting higher redshift objects. The hierarchy of redshifts in the various associations also confirms the working model of larger older galaxies bringing forth the original ejections, and then the younger, higher redshift offspring giving rise to even higher redshift ejections. The galaxy clusters seem to be able to eject quasars up to about z = .6. But by about z = .3 the progenitor quasars and BL Lac objects seem to more easily break up into smaller galaxies and form clusters. Both the quasars and these galaxy clusters, however, are much closer, smaller, less luminous, less massive and much younger than conventionally assumed.

The Archetypal Coma Cluster of Galaxies

The Coma Cluster represents the densest, most conspicuous aggregate of galaxies in the sky and has long been taken as the prototype, medium distant, galaxy cluster. Sinclair Smith was one of the first astronomers to calculate, under the assumption that the redshifts were velocities in equilibrium (virialized), that the mass of the cluster far exceeded the mass of the galaxies comprising it. Fritz Zwicky later emphasized this discrepancy and so was born the concept of "dark" or "missing mass." This observationally undetected, but crucial repair to the theory had to be invented to save the redshift = velocity assumption. Eventually it became so needed that today we have a universe which is reported to be about 90% unobservable.

The first thing one should notice about the Cluster, however, is that the brightest galaxy has a redshift of z = 6456 km/sec whereas the mean of the rest of the galaxies has z = 7000 km/sec. This is an extreme case of the companions having systematically higher redshift and therefore a considerable component of non-velocity redshift. Bye

bye dark matter! (Also the intrinsically redshifted galaxies in young, non-equilibrium clusters—*e.g.* X-ray clusters—become much less massive and therefore less able to lens background objects and more like the low redshift quasars and BL Lac objects to which they were related in Chapter 6.)

But the X-ray map of the cluster shown in Figure 7-19 is a real shocker. Far from an equilibrium distribution of X-rays, the cluster is extended with strong X-ray sources on either end of the extension. The strong source on the SW extension (note the isophotal elongation toward it) is not identified. But the strong source to the NE is an active galaxy (HII) with a redshift of z = .029 (8700 km/sec). In this same direction lies the strongest point X-ray source within the area of the cluster, X Comae, with a redshift of z = .092. Why should an active galaxy bright enough to be named in the variable star catalogue fall right in the Coma Cluster? Why should these events always be explained as accidental background objects?

But then why should the Coma Cluster have always been assumed to be a prototype cluster when it is a unique aggregation of lenticular and E galaxies without the usual complement of other morphological types to give some hint of their luminosity? As Zwicky sagely remarked, "It takes a large number of stars to make a spiral galaxy, but only three to make an E galaxy." Since we do not have any redshift or morphological criterion for its distance, the best evidence seems to be its location in the sky, where it forms an obvious northern continuation of the Virgo Cluster in the Super Galactic plane.

X Com is a 16.65 mag. Seyfert 1 galaxy much like those found in the Virgo Cluster (*e.g.* PG1211 +143) and looking around it furnishes some stunning proof of the association of quasars with Seyferts. As Figure 7-19 shows, *almost all the point X-ray sources are clustered around X Com.* They are catalogued quasars with their redshifts written next to them in the Figure. The redshifts are in the expected range and they form a rough line through the Seyfert galaxy. It would be very interesting to identify and measure the X-ray source just to the NW of X Com along this line.

Altogether the Coma Cluster confirms in general and in detail what we have learned about clusters of galaxies. Instead of being old, quiescent systems they are filled with high-energy radiation that requires resupply. They show strong evidence for recent ejection of higher redshift matter. They have large components of intrinsic redshift. They are small, and low luminosity and are associated with older, more nearby galaxies.

The Hubble Deep Field

In a courageous assignment of discretionary time, the director of the Space Telescope Science Institute, Robert Williams, assigned 150 orbits of the space telescope to deep photography of a single 4 square arc minute field at R.A. = 12h 36m 49s, Dec. = +62d 12' 58" (position calculated for equinox 2000.0). This reached objects of far fainter apparent magnitude than had ever been seen before, about R = 30 mag. The resulting picture is shown here in Plate 7-20.

There are obviously a very large number of very peculiar and unusual galaxies in this field. But even after examining over 77,000 galaxies in the *Arp/Madore Catalogue* of southern Peculiar Galaxies and associations, I did not appreciate just how peculiar the galaxies in the deep field were until I performed a galaxy by galaxy classification on the same system of the previous Catalogue of bright galaxies. Table 7-1 here shows that while 95% of the nearby galaxies have normal, regular morphologies, only 11% of the Deep Field galaxies could be considered normal in appearance. (I sent this report to Nature, but even before John Maddox left the editorship, the magazine had elevated itself above simple observational results and it was not published). My friend and classification expert, Sydney van den Bergh, added another important result, namely that there were almost no normal, grand design spirals in the deep field.

As might be expected, some of the redshifts measured are very high going into the range of z = 3 to 4. What does this mean in terms of what we have found out about the nature of redshifts? It will be shown in the theory chapter upcoming that our interpretation of the redshifts agrees with the conventional interpretation in that we both agree that a high redshift object is very young. (In the conventional view they are far away and the long look back time shows how they looked when they were much younger. In our interpretation they are the same young age—recently created—but can be quite nearby.) The essential difference is really that our newly created objects are low luminosity and the Big Bang young objects are very luminous because they must be seen at a great distance. So the question comes down to: "Are the young objects seen in the Deep Field low luminosity or high luminosity?

Empirically the highest luminosity objects we know without recourse to redshift distances are the regular, Sb spirals. We would generally expect the most luminous objects to be the most massive and therefore the most relaxed, equilibrium forms. This is one thing the Hubble Deep Field objects are not. On the other hand the quasars we interpret as young, low luminosity objects tend to be high surface brightness, small apparent angular diameter objects. Would even lower luminosity quasars at high redshifts look like the Hubble Deep Field objects? Would 3C48 and the quasars examined for host galaxies with the space telescope be comparable? That would be an interesting point to discuss in detail. But one remark in this direction is that the tendency for young, nearby, low luminosity objects to break up, eject material, show jets and disturbances could explain the prevalence of linear, knotty objects and multiple objects as observed in the Hubble Deep Field.

Table 7-1 Hubble Deep Field compared to Nearby Galaxies

	Irregular	Irregular Sp.	I/a gals.	DBL	LSB	Normal
Deep field number	59	27	87	25	8	51
Deep field percent	57%	06%	19%	05%	02%	11%
Nearby percent	1.4%	.5%	2.2%	.9%	.5%	94.5%

(I/a = interacting, DBL = Double, LSB = low surface brightness)

Richard Ellis has studied the Hubble Deep Field in detail. He finds a large number of a new kind of low surface brightness, peculiar galaxy. On the conventional interpretation of redshift, they are of high enough luminosity that it is clear that there can be no local counterparts of such objects. But isn't it required then that they be nearby, low luminosity remnants of more recent formation which have higher intrinsic redshifts because they are young?

This is an important point because our evidence has been going in the direction that all objects we can be sure of are within the rough confines of the Local Super Cluster. But are there any objects appreciably beyond the Local Super Cluster or is there empty space for an unknown distance? On the interpretation of redshift this book uses, there could be very distant objects of any redshift (age), but they would have to be luminous enough so that we could see them at great distances. If there is an upper limit to galaxy luminosity we may not have seen any so far. If we go fainter as with Space Telescope, we may see some if they are there, but how will we recognize them amongst the welter of nearby objects?

An Amateur Spots the Crucial Patterns

The empirical pattern recognition which has so drastically changed our view of extragalactic astronomy in this book is based on the recurrent evidence of pairing of active objects across large low redshift galaxies. The tyranny of the observations is to insist on opposite ejection of extragalactic material as a ubiquitous process that operates on all scales. How is it possible that the exquisitely trained professional scientists have not recognized this evidence?

To make the point that it is not the evidence but the viewer that is the key here, I want to present Figure 7-20. These are examples from a page which an architect named Leo Vuyk sent me. He just Xeroxed out the pictures which he found in astronomical publications which showed the same theme over and over again. He drew in the lines— like the delightful childhood game of connecting the dots together to get the picture. None of us have the correct theory, but the professional tends to interpret the pictures by using the theory he was taught while the amateur tries to use the picture to arrive at a theory.

The Real Life Story of the Exploding Radio Galaxy 3C227

I sat down at lunch across from a long time acquaintance of mine from a neighboring institute.

"Oh Chip", he said, "you will be interested in an object in the midst of a disturbed galaxy I am analyzing."

A few weeks later he said: "There is an emission line which seems to indicate the object has a lower redshift than the radio galaxy."

I replied, "I cannot understand that, I would expect it to be higher."

Many months later he reported: "Oh, that object turned out to have a higher redshift."

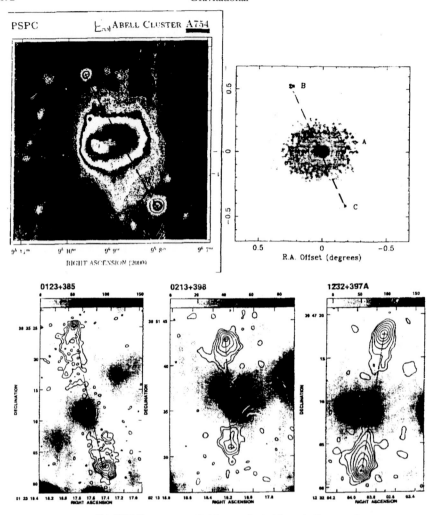

Fig. 7-20. From a page of pictures collected from the literature by the architect Leo Vuyk. He has drawn in the lines connecting the companion objects across the central object.

For years after that every time I passed him on the grounds I urged, "Bob, you should publish that observation, it is very important."

After a long time I had finally forgotten about it when a reprint appeared on my desk (*Mon. Not. Roy. Astr. Soc.* 263,10,1993). I remember thinking, "Is this the object"? Rapidly I scanned the pictures trying to spot the object. No luck. Finally I located a small diagram on which an object was marked "BO." "What is BO?" I wondered. I could only think of an advertising jingle from my childhood, "Lifeboy soap fights B.O. (body odor)." Then I found at the end of the figure caption, "BO indicates a background object of redshift z = .3799."

It took some time for me to figure out where it was on the picture of the disturbed radio galaxy (which had a redshift of z = .086). It was amazing, it was only 11

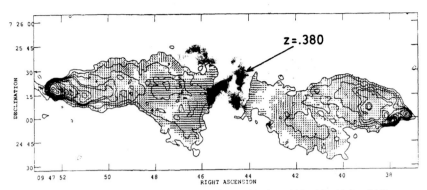

Fig. 7-21. The violently disrupted radio galaxy 3C227 (z = .086) with a high redshift object of z = .380 only 11 arc sec from the center. Dark features in interior show emission from excited gas and contours show radio emitting material ejected from the center of this Seyfert 1. Figure adapted from M.A. Prieto *et al.*

arc sec from the center of the explosion! *That meant the chance of finding a background quasar was only about 3 in 100,000!* On the other side, and further out, were some compact galaxies of about z = .129. It was a more or less perfect example of an ejection into the dense part of the galaxy which had ripped the galaxy apart and slowed down the escape of the high redshift object (as suggested in Chapter 3, *e.g.* Figure 3-29 and 3-30). Would shells be formed further out due to the explosion? There was no sign of arc deformation of the high redshift object or any other evidence for gravitational lensing. *As a final, finishing touch I looked it up in a Catalogue and found the central galaxy was a Seyfert 1!*

It was only then that I noticed the last sentence of the paper which said:

> *Finally we note the discovery of an object with z = .3799... Its alignment with one of the brightest extranuclear regions in 3C227 is remarkable.*

After 27 years of evidence for the physical association of such objects I would love to know what went through the authors' minds when they decided to use, instead of the neutral term "high redshift object", the term "background object."

Chapter 8

QUANTIZATION OF REDSHIFTS

The fact that measured values of redshift do not vary continuously but come in steps—certain preferred values—is so unexpected that conventional astronomy has never been able to accept it, in spite of the overwhelming observational evidence. Their problem is simply that if redshifts measure radial components of velocities, then galaxy velocities can be pointed at any angle to us, hence their redshifts must be continuously distributed. For supposed recession velocities of quasars, to measure equal steps in all directions in the sky means we are at the center of a series of explosions. This is an anti-Copernican embarrassment. So a simple glance at the evidence discussed in this Chapter shows that extragalactic redshifts, in general, cannot be velocities. Hence the whole foundation of extragalactic astronomy and Big Bang theory is swept away.

The early history of the 72 km/sec redshift periodicity in galaxy redshifts is discussed in *Quasars, Redshifts and Controversies* (starting on page 112). Subsequent investigations have confirmed that period and increasingly accurate redshift measures have established another even more conspicuous period of 37.5 km/sec. Meanwhile the larger redshifts of the quasars have received added support for their periodicity from increased numbers of measured redshifts and also the beginning of the ability to correct certain pairs for ejection velocity components. Narrow beam surveys of field galaxies have shocked establishment astronomers by showing periodicity. (The shock would have been less if they had not disregarded 20 years' worth of previous observations of redshift periodicities.)

On the theoretical front it has become more persuasive that particle masses determine intrinsic redshifts and that these change with cosmic age. Therefore episodic creation of matter will imprint redshift steps on objects created at different epochs. In

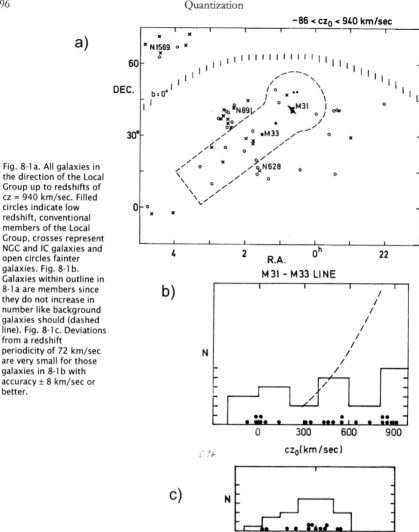

a) $-86 < cz_0 < 940$ km/sec

Fig. 8-1a. All galaxies in
the direction of the Local
Group up to redshifts of
cz = 940 km/sec. Filled
circles indicate low
redshift, conventional
members of the Local
Group, crosses represent
NGC and IC galaxies and
open circles fainter
galaxies. Fig. 8-1b.
Galaxies within outline in
8-1a are members since
they do not increase in
number like background
galaxies should (dashed
line). Fig. 8-1c. Deviations
from a redshift
periodicity of 72 km/sec
are very small for those
galaxies in 8-1b with
accuracy ± 8 km/sec or
better.

addition it appears increasingly useful to view particle masses to be communicated by
wave like carriers in a Machian universe. Therefore the possibility of beat frequencies,
harmonics, interference and evolution through resonant states is opened up.

The 72 km/sec Periodicity

I will only comment on two of the more recent observations of this particular
periodicity. The first involves periodicities of redshifts which are members of our Local
Group of galaxies. In *Quasars, Redshifts and Controversies*, Figures 8-11 and 8-12 show that
there are a number of smaller galaxies and hydrogen clouds that are distributed along
the minor axis of M31, implying that the central galaxy in our Local Group is ejecting or
has ejected material in this direction. The first figure here, 8-1a, shows that galaxies up

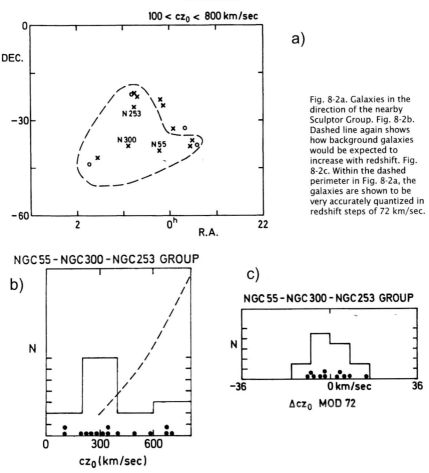

Fig. 8-2a. Galaxies in the direction of the nearby Sculptor Group. Fig. 8-2b. Dashed line again shows how background galaxies would be expected to increase with redshift. Fig. 8-2c. Within the dashed perimeter in Fig. 8-2a, the galaxies are shown to be very accurately quantized in redshift steps of 72 km/sec.

to $cz = 940$ km/sec redshift belong to this line and are therefore younger members of the Local Group. Figure 8-1b shows they are not background galaxies which would increase in numbers sharply with fainter apparent magnitudes. (Figure 3-16 in a previous chapter showed that this is a typical range for accepted members in more distant groups). The point of all this for quantization of redshifts is that these Local Group companions demonstrate strong periodicity at 72 km/sec as shown in Figure 8-1c. The details of this study are presented in the *Journal of Astrophysics Astron.* (India) 8, 241, 1987.

In Figure 8-2 the same analysis is performed on the Sculptor Group which is a small group of galaxies between the Local Group and the next major, M81 group. In these two nearest groups, faint galaxies with accurate redshifts have been uniformly surveyed and their redshift distribution clearly shows the 72 km/sec period. Figures 8-1c and 8-2c show that even though the redshifts reach up to a multiple 14 times 72.4 km/sec, the average deviation from the period is only ± 8 km/sec. But this is just the average measuring accuracy of the redshifts! In fact for the 7 redshifts which are known with greater accuracy, the periodicity fits within about 3 or 4 km/sec.

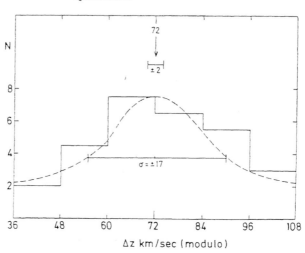

Fig. 8-3. All accepted members of the Local Group and M81 group as shown in Chap 3, Fig. 3-2 are here analyzed for periodicity in redshift. The period comes out 72.4 ± 2 km/sec with mean dispersion of about 17 km/sec.

Another development to note is that the original Tifft determination of the period was 72.46 km/sec. In a subsequent analysis of the Local Group and the M81 group with a more accurate value of the solar motion correction than commonly used, Arp found a value of 72±2 km/sec (Figure 8-3). This was followed by the more extensive investigation of the Local Group described above in which the large number of multiples enabled the periodicity to be again found to three significant figures as 72.4 km/sec. It is impressive that such exact numerical agreements have resulted.

It is also perhaps revealing to comment on the latest result concerning the 72 km/sec periodicity. In the Tucson conference on quantization in April 1996, one of the long time opponents of quantization presented a small number of new measures on double galaxies.

"You see," he said "These galaxies which are at greater separation show a 72 km/sec periodicity, but when the galaxies get closer together where the pairs are more reliable—the periodicity disappears."

He beamed out at a momentarily silent audience. Perhaps because I have more experience in these matters now, I was the first to put up my hand and reply:

"When the galaxies get closer together their orbital velocity about each other increases and smears out the quantized redshift steps. It's just what you would expect."

Nothing else was said but the next day he got up and said:

"But I just showed yesterday that the observations contradicted the 72 km/sec quantization."

So I had to stand up and repeat what I had said the day before. Fortunately the meeting ended after that.

The 37.5 km/sec Periodicity

One pair of researchers who seriously tested the quantization (with the initial expectation of disproving it) was Bruce Guthrie and William Napier, both then at the Royal Observatory at Edinburgh. They developed and applied especially rigorous

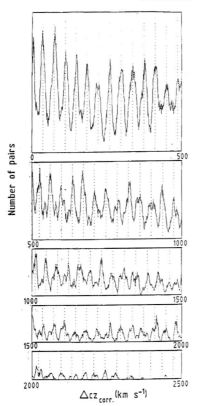

Fig. 8-4. From an analysis by Bruce Guthrie and William Napier which shows that all the most accurately known galaxy redshifts out to about 2500 km/sec are accurately quantized in steps of 37.5 km/sec.

statistical tests to galaxies with accurate redshifts in the direction of the Virgo Cluster. They found the galaxies in the outer regions to be quantized in 72 km/sec steps but not the inner. It was obvious that just as in the above-related Tucson incident, the inner parts of the Virgo Cluster, deeper in the potential well, were moving fast enough to wash out the periodicity. It is nice to see something work as expected in this business.

But in the process of this and later testing, Napier and Guthrie, who were working with the best enlarged sample of most accurate redshifts, saw an extremely prominent periodicity at 37.5 km/sec. Figure 8-4 shows their most recent analysis of the most accurate set of Hydrogen line redshifts. One can see at a glance how accurately the troughs and peaks of redshift march metronomically outward from 0 to over 2000 km/sec. It is typical science protocol that such obvious results have to be tested for numerical probabilities. Figure 8-5 shows the result of a Fourier analysis which picks out the 37.5 km/sec period. Overall the significance of the effect is one in a million.

This result first appeared in detail in *Progress in New Cosmologies* (Plenum Press). Then it took four years to struggle through science's most treasured institution, peer review. (The referee demanded an analysis of a whole new group of redshifts, which as it turned out, confirmed all the previous results). But *Science* magazine (18 Dec. 1992, p.1884) reported the first result in a small news note in which they quoted Joe Silk, well

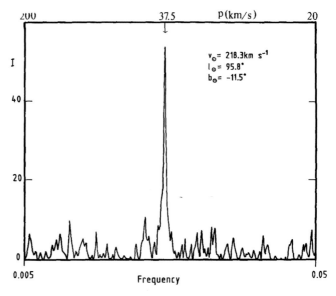

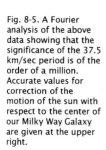

Fig. 8-5. A Fourier analysis of the above data showing that the significance of the 37.5 km/sec period is of the order of a million. Accurate values for correction of the motion of the sun with respect to the center of our Milky Way Galaxy are given at the upper right.

known theoretician at the University of California, Berkeley as saying: "It's just noisy data." James Gunn of Princeton added: "We have a lot of crank science in our field."

Four years later *Science* (9 Feb 1996, p759) reported on it again, quoting John Huchra of Harvard as saying: "I am thinking of writing a proposal for checking to see if [the effect] holds up with other galaxies." Since Napier and Guthrie had used up the most accurate neutral hydrogen redshifts of accuracy better than 4 km/sec for their most significant results, any new redshifts that were optical absorption line redshifts would have lesser accuracy (with redshifts systematically 20 km/sec less). That could only give a lesser confirmation which would undoubtedly be accompanied with the news "Oh the effect is going away with a larger sample." But just so that people wouldn't think that *Science* (the publication) had gone soft on crank science, they finished with a quote from James Peebles, Princeton cosmologist: "I'm not being dogmatic and saying it cannot happen, but..."

Machian Physics?

As an illustration of the difficulty of finding an acceptable solution for the riddle of the 37.5 km/sec periodicity, I outline the following proposal:

The luminosity weighted mean of the Virgo Cluster is 863 km/sec (from Chapter 5). The observations of redshift periodicity would require that 23.0 steps of 37.5 km/sec were required to go from the redshift of our own galaxy (MW = 0 km/sec) to Virgo (V = 863 km/sec). *It is very unlikely that the galaxies between us and the distance of Virgo are distributed in 23 shells centered on our galaxy.* Moreover the peculiar velocities of the galaxies must be less than about 20 km/sec in order not to wash out the periodicity. One can see immediately the difficulty of finding a reasonable solution.

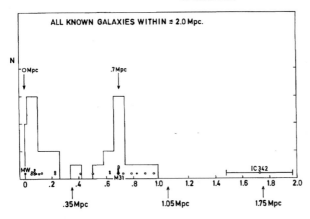

Fig. 8-6. All known galaxies within about 2.0 Mpc of our own galaxy are shown. Empty regions appear to occur at .35 and 1.05 Mpc. The distance to last galaxy, IC342, is estimated to be uncertain by the appended error bar.

But suppose we have quiet galaxies and we look for a reason why we do not see galaxies between these redshift steps. If we adopt the premise that elementary particles acquire mass by exchanging machions (the signal carrier for inertial mass) with other matter within their light travel horizon, then we should ask what is the De Broglie wavelength of this machion. (I call this a machion in analogy with the supposed carrier for gravitational mass which is commonly called the graviton). The point is that when matter is at a distance from us where the oscillating wave is 180 degrees out of phase with our own matter, then we will not know about that matter. The photons emitted by that matter will not know about our detectors and will not be absorbed by them. This is in accordance with the resonance theory of the structure of mass particles by Milo Wolff.

If there are 23 machion wave maxima between us and the Virgo Cluster which is at a light travel distance of 1.6×10^{15} sec, then the frequency of the machion must be $1.4 \times 10^{-14} s^{-1}$. The Compton frequency of the electron is $1.2 \times 10^{20} s^{-1}$ and therefore the mass of the machion $m_m = m_e \times 1.2 \times 10^{-34}$. That is the mass of the machion would be 34 orders of magnitude smaller than the mass of the electron. This would still, however, be about 200 times the mass of the "soft boson" which Hill, Steinhardt and Turner investigated as a possible origin for the pencil beam periodicity.

If we adopt the distance to Virgo as 16 Mpc then the apparently empty zones would be spaced about 0.7 Mpc apart. That would allow us to see everything out to about 0.3 Mpc before we encountered a zone of appreciable fading. M31 at a distance of 0.7 Mpc would be at a point of maximum visibility.

The principle advantage of this solution is that it would allow a continuous distribution of galaxies in space and not require concentric shells of galaxies centered on the observer. The solution would work equally well for a smoothly expanding universe as well as a non expanding universe with redshift caused by increasing look back time with increasing distances.

It is possible to test the predictions of the above mechanism in a way that no other model is capable of being tested. Because all the galaxies in our Local Group are almost certainly known out to about a radius of about 2.0 Mpc, it is possible to plot their distances as done in Figure 8-6. The census of galaxies and their distances are taken from the work of Bruno Binggeli.

The surprising result is that of the 26 galaxies plotted, they almost all concentrate at two distances separated by about 0.7 Mpc. Or, perhaps more significantly there are troughs vacant of galaxies at about .35 and 1.05 Mpc from our own galaxy. It is true, of course, that there is a sub-condensation of companions around M31, but the remainder of the galaxies are seen in various directions around the sky. M31 as the center of our Local Group would be expected to have its companions spread out within a radius of about 1 Mpc as is common in most groups. The distance to the first peak of galaxies did not have to come out at .7 Mpc and there did not have to be, and in fact would not be expected to be, a clear falling off of galaxies on either side.

The furthermost member of the Local Group, IC342, has an insufficiently accurate distance to decide whether it falls at a peak or a valley. In fact this illustrates that the machion explanation advanced here can not be tested by plotting more distant galaxies because the absolute accuracy of their distance determinations is not sufficient.

My attitude toward this result is that in a Machian universe there must be some signal carrier for inertial mass coming from distant galaxies. We would expect this machion to be small compared with the mass of the electron. Therefore, if the most conspicuous and unexplained redshift anomaly is explained by this concept, we should perhaps consider that astronomy has made a measurement of a physical quantity far below the possibility of terrestrial physics laboratories.

The surprising departure of this proposal from our normal assumptions serves to illustrate the difficulty of explaining the observations conventionally. It invites further complexities such as interfering frequencies and beat frequencies. As we shall comment later, however, it probably does not explain the 72 km/sec periodicity of redshifts. We can see already that 72.4 km/sec is not, within the errors, twice the 37.5 km/sec periodicity.

Pencil Beam Surveys of Galaxies

What little attention was directed toward the 37.5 and 72 km/sec periodicity was disparaging, along the lines of "it is obviously just incompetent observers." But in 1990 some respectable astronomers measured many galaxies in a small field and found clumping of redshifts. The investigators, after considerable delay, rather nervously announced this result, but only after turning the redshifts into distances via the obligatory redshift-distance relation. To obtain the primary data one had to read off the preferred redshifts from their graphs. It is clear their main peaks were around z = .06 and .30 with some fine structure.

This is extraordinarily interesting because this coincides with the first two peaks of the redshifts for quasars. In fact this is a reassuring confirmation because, after all, quasars are an active form of galaxies—or put another way—it was demonstrated in 1968 that there was a continuity of physical characteristics between quasars and galaxies.

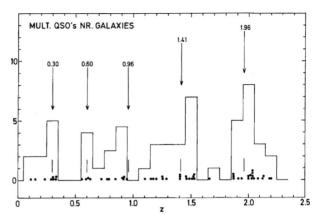

Fig. 8-7. Distribution of redshifts of quasars where more than one is associated with a low redshift galaxy. Peaks predicted from Karlsson's formula are marked.

The pencil beam surveys were timidly ventured as sheets of galaxies periodically spaced at distances of about 128 Mpc. This produced some tentative models for their origin involving scalar fields coupled by very weakly interacting particles and even some coupling to curved space time models. It is interesting that the relativistic models were conformal transformations of the Einstein field equations—as is the Narlikar/Arp, redshift-as-a-function-of-time solution. Perhaps there is some common ground if the particle physics language and the geometrical language can be translated.

Quantized Quasar Redshifts

In 1967 Geoffrey and Margaret Burbidge pointed out the existence of some redshifts in quasars which seemed to be preferred (particularly z = 1.95). In 1971 K.G. Karlsson showed that these, and later observed redshifts, obeyed the mathematical formula $(1+z_2)/(1+z_1) = 1.23$ (where z_2 is next higher redshift from z_1). This gives the observed quasar redshift periodicities of: z = .061, .30, .60, .91, 1.41, 1.96, *etc.* In my opinion this is one of the truly great discoveries in cosmic physics. He was rewarded with a teaching post in secondary school and then went into medicine.

Many investigations confirmed the accuracy of this periodicity. And of course, many claimed it was false. One postdoctoral student at the Institute of Theoretical Astronomy in Cambridge, where Martin Rees was Director, claimed there was no periodicity. His analysis included the faintest, least accurate quasars which had been shown *not* to exhibit periodicity. They showed it anyway. In a *new* sample of X-ray quasars, he found the periodicity but issued the opinion that it would go away with further measures (fainter quasars). We will see the opposite happened.

A Wise traveler from the East

The astronomer Y. Chu from Hefei, China walked into my office in the Max-Planck Institut one day. He said, "I find that the quasars that you associated with low redshift companion galaxies (starting from 1967) exhibit the redshift periodicity particularly well." I carefully made out the complete list of associations known at that time. Figure 8-7 here shows the distribution of redshifts of quasars in the most secure cases where more than one is associated with a single low redshift galaxy. It is striking how the redshift peaks predicted by Karlsson's formula fit the observed distribution for this group of quasars never before tested.

A young Chinese doctoral student at the Max-Planck Institut für Astrophysik named H.G. Bi started analyzing the periodicity with power spectrum analysis and together with Chu and his wife, X. Zhu, we undertook a thorough investigation of all the data available. For multiple quasars near galaxies we found that the predicted periodicities were fit by the formula at the 94% confidence level. If we made the small correction for the redshift of the parent galaxy, the confidence level increased to 95%. If we omitted one of the 14 groups which was discordant, the confidence level rose to 99.5%.

Now one of the ongoing attempts to discredit the redshift periodicity was an argument that quasars were discovered by their ultraviolet excess and that excess was caused by prominent emission lines moving into the ultraviolet window at certain redshifts—in other words the periodicity was merely a selection effect. It had been shown that this was not the case, but nevertheless the argument was widely accepted as disproving this embarrassing observational result. In order to try to set this to rest once and for all, we selected quasars that had only been discovered by their radio emission.

The Right Ascension = 0 hour and 12 hour region are the two principle regions in which we can avoid the obscuring plane of our own galaxy in looking out into the extragalactic sky. We divided the radio quasars into the 0 hour group and the 12 hour groups and only accepted quasars with radio strengths greater than 1 flux unit. Figure 8-8 shows the results—the strongest confirmation yet obtained!

It is noticeable that there is a small, 3%, offset between the zero point of the periodicity in the 0 hour region and that in the 12 hour region. This represents a small difference between the redshifts of the quasars in the direction of the Local Group and those in the direction of the center of the Local Supercluster. I continue to feel that this represents an important clue both to the cause of the periodicity and the structure of the Local Supercluster.

But in establishing the reality of the periodicity the results are overwhelming. Table 8-1 shown here is abstracted from *Astronomy and Astrophysics*. 239, 33, 1990. It shows the 0 hour and 12 hour regions separately confirm the period with 99 and 96% confidence limits. Together they confirm at 99.97% confidence. If we make the 3% shift on zero point before we add the two samples, the confidence is 99.997% or only one chance in about 33,000 of being accidental.

One point should be strongly emphasized: In the *A&A* paper it is stated "We should note that plotting all quasars listed in *Hewitt and Burbidge* [catalogue] with z > 1.3 in these 0 hour and 12 hour regions down to the faintest apparent magnitudes shows no conspicuous periodicity." In other words the periodicity becomes less pronounced at fainter magnitudes. Since redshift is an intrinsic property and not a measure of distance, at a given redshift the best indication of large distance is a faint apparent magnitude.

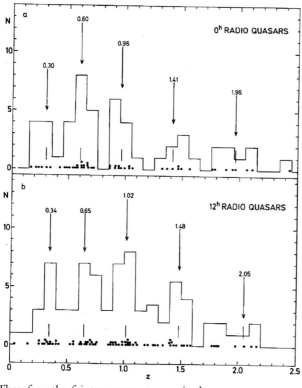

Fig. 8-8. Catalogued quasars found by radio emission with greater than 1 Jansky flux at 11 cm. Quasars in Local Group Direction (0^h) and Local Supercluster direction (12^h). Latter are shifted by $(1+z) \times (1 + 0.03)$.

Therefore the fainter apparent magnitude quasars are more distant and we are probably seeing a change in periodicity with distance.

The bright apparent magnitude, high redshift (around $z = 2$) quasars are mostly at the relatively close distance of our Local Group (See Distribution of Quasars in Space, Chapter 5, *Quasars, Redshifts and Controversies*.). We can see lower redshift quasars out to the distance of the Local Supercluster. But if there is not much beyond the boundaries of the Local Supercluster then we should see the quasars of $.5 < z < 1.0$ becoming

Table 8-1 Probablity of Periodicity being Accidental

Sample	No.	Prob. Accident	Remarks
Multiple quasars near galaxies			
Near comp^n's	54	.061	Associations as of 1987 (see Fig. 8-7)
" "	54	.049	Allowance for redshift of central gal.
" "	49	.005	omitting NGC 2916
Radio selected quasars			
0^h	50	.013	See Fig. 8-8
12^h	73	.039	" "
$0^h + 12^h$	123	.0003	Combined 0^h and 12^h regions
$0^h + 12^h$	121	.00003	12^h group shifted by .03

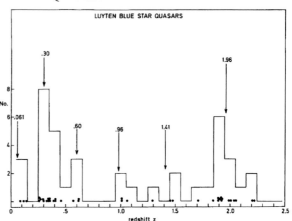

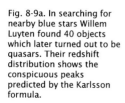

Fig. 8-9a. In searching for nearby blue stars Willem Luyten found 40 objects which later turned out to be quasars. Their redshift distribution shows the conspicuous peaks predicted by the Karlsson formula.

relatively less numerous at fainter apparent magnitudes. The observations confirm this expectation as shown in Figure 7-2 of the gravitational lens chapter.

The Closest Quasars

In the 1940's and 1950's Willem Luyten measured blue stars looking for large proper motions which would identify nearby blue dwarf stars. After all this time it turns out that 40 of his stars are quasars. Figure 8-9a shows the distribution of their now known redshifts. They conspicuously outline all the major quasar redshift peaks.

It is interesting to consider that being measured so many years ago, they are rather bright apparent magnitudes and therefore probably represent the nearest quasars to us.* But the strongest peaks in Figure 8-9a support the conclusion that the quasars in the $z = .30$ and 1.96 peak are generally the lowest luminosity and thus are seen in relatively greater numbers nearby. Compare the fainter apparent magnitude radio quasars in Figure 8-8 to see that the peaks of $z = .60, .96$ and 1.41 are much stronger, agreeing with the previous conclusion that they are the most luminous quasars that can be seen at greater distances (for example in the 12h direction toward the Local Supercluster).

As mentioned previously, it has been argued in the past that strong emission lines in the spectra can cause certain redshifts to be favored for selection as blue objects. But actual analysis shows the emission lines in general are not strong enough to cause this effect. Moreover, we have just shown that quasars selected for their radio properties, not their color, show the redshift peaks very clearly. Moreover, the BL Lac objects, a kind of quasar with the typical quasar blue continuum, but with negligible emission lines, shows the same redshift quantization (with emphasis, as seen below, on the lower redshift peaks which we have associated with relatively nearby objects).

Periodicity of BL Lac Redshifts

* At the distance of the Virgo Cluster (16 Mpc) a quasar travelling at .1c (30,000 km/sec) would have a proper motion of .4 milliarcsec per year. At the distance of M31 (.7Mpc) the motion would be 9 milliarcsec per year. The errors quoted for the Luyten proper motions are about ± 18 milliarcsec per year. Nevertheless the Luyten list should be examined for possible significant proper motions as J. Talbot has started to do.

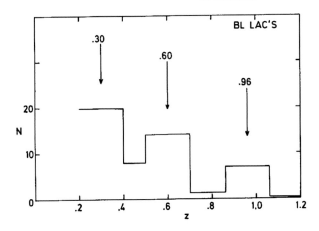

Fig. 8-9b. All catalogued BL Lac objects (Véron and Véron 1995). Bins are ± .1 z with three major quasar redshift peaks marked.

Since it has been emphasized throughout that BL Lac objects are a kind of quasar, they should show the same redshift periodicity as the quasars. Although they are much rarer, their strong X-ray and radio emission enables a fairly complete sample of the brightest objects in the class to be catalogued. I finally got tired of seeing results presented on these objects where the investigator was oblivious to the fact that they were at the quasar quantized redshift values. So I went to Table 2 of the Véron and Véron Catalogue and plotted the redshifts of all the known BL Lac objects. *Figure 8-9b shows they have exactly the quasar peaks at z = .30, .60 and .96.*

There is a very interesting aspect of this plot, namely that I have binned in ± .1 z intervals. This is the amount we would expect the redshifts to be spread around their intrinsic value due to average ejection velocities of .1c (as found in Chapters 1 and 2). This nicely encompasses the observational spread, the centers of which are closely at the quasar values. People who are acquainted with the quasar periodicity will know, however, the formula predicts the shortest period at z = .061. This peak is confirmed by the observations (*A&A* 239,33). The BL Lac's have their strongest numbers in this range, but in Figure 8-9b the 0.1z spread overlaps them with the low redshift side of the z = .30 peak, so the BL Lac periodicities are merged in this range in the graph.

This z = .06 peak for the BL Lac's is very important because as we recall in Chapter 6 the X-ray clusters have a very sharp peak in their redshift distribution at exactly this redshift (Figure 6-17). We argued then that the galaxy clusters came from breaking up of BL Lac objects and to find a matching redshift distribution for them is strong support for this radical idea!

One point that I have noticed and which may have also impressed others is that the BL Lac's found associated with Seyfert galaxies in Chapter 2 generally had very exact correspondences with the quasar periodicity peaks. This was also true of the low redshift quasars around z = .30 which seemed like BL Lac's in a low excitation state. This might be reconciled with the 0.1c ejection velocities and the similar spread around the BL Lac peaks found in Figure 8-9 by supposing that there are those ejected quasars which escape into the field and those which are captured by the ejecting galaxy. Those which are captured, that is associated with the parent galaxy, have necessarily slowed

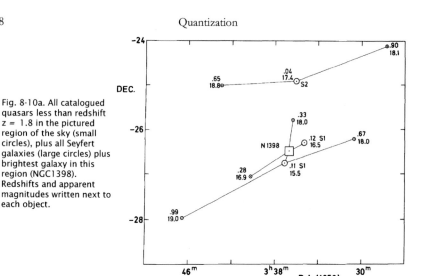

Fig. 8-10a. All catalogued quasars less than redshift z = 1.8 in the pictured region of the sky (small circles), plus all Seyfert galaxies (large circles) plus brightest galaxy in this region (NGC1398). Redshifts and apparent magnitudes written next to each object.

their velocities. The velocity spread from the intrinsic redshift peak would then be much smaller. Since the quasars evolve to lower redshift with time, the lower redshift ejecta around z = .30 would have been traveling longer and had more time for the Narlikar/Das mechanism to slow them down (Chapters 2 and 9).

That the quasars physically associated with their galaxies of origin had slowed down their ejection velocities would also nicely explain why the quasars initially found around companion galaxies had particularly well defined periodicity.

Periodicity Patterns on the Sky

Really all we have for data in astronomy is photons as a function of x and y and frequency. The challenging puzzle is then to try to reason out how nature works. I think this is best done by pattern recognition—what is related to what—and in what recognizable way. As a kind of test of what I think I have learned, I present Figure 8-10a here. It is a region of the sky I just happened upon. This implies that there are many more regions like this and that anyone can play this game with a pencil, graph paper and catalogues of extragalactic objects. In fact I think it is better that other people inspect the catalogues and present the results. If others do it there may be a better chance of having these exceedingly important results accepted and utilized.

One can start anywhere. I started by plotting all the quasars less than about redshift z = 1.5. There seemed to be three pairs. Then I asked where are the Seyfert galaxies that gave rise to these pairs. I found two listed under the catalogue of Seyferts and one listed as a quasar. (Another falls just at the SE corner of Figure 8-10a but I ignore that one.) But then the most electrifying thing happens. One of the Seyferts falls closely between the upper pair of quasars and another falls between the lower pair of quasars. These are the only catalogued Seyferts in the area!

At this point I say, well the active Seyferts that eject quasars are usually companions to (have been originally ejected by) some nearby, large, low redshift galaxy.

So I plot the brightest galaxy in the pictured region. It falls very close to, and just between the two lower Seyferts! It is a m = 10.6 mag. SBab, just the general class of galaxy one would expect to find at the origin of a group of galaxies of different ages. And why, over the whole region that it could have appeared by accident, did it appear at just the focus of activity?

But the fun is just beginning. Notice the redshifts of the objects involved. The right hand member of the upper pair at z = .90 is .06z less than the quantized peak of .96 and the left hand member at z = .65 is .05z more than the quantized peak at .60. It is almost exactly as if the right one had been ejected towards us and the left away from us. Dropping down to the pair at z = .33 and .28 it is as if the right hand one was traveling away from us at .03z and the left hand one toward us at .02z—with the intrinsic redshift almost exactly at the z = .30 quantization peak. The third pair does not quite work out, with one going away at .07z and one at .03z. But consider the mean redshifts of the pairs of quasars (which averages out the toward and away ejection velocities). The formula peaks are on the left, the observed pair averages on the right:

z_n	z_{ave}	
.06	.09	(3 Seyferts)
.30	.305	(2 QSO's)
.60	.66	"
.96	.945	"

Can this be an accident? If it is not an accident, the whole basis of modern extragalactic astronomy falls. What will the professionals decide? What will individuals decide?

How close we should expect an individual quasar to fall to a quantization peak probably depends on the details of its evolution. The newly ejected object should start out with a high redshift and evolve to lower redshift as it ages and becomes more luminous. In order to exhibit steps in redshift it must evolve more rapidly between redshift peaks which might be viewed as resonant states. But we do not know how broad the states are or what the chance is of catching a given object between states.

The final interpretation of Figure 8-10a is not entirely fixed, however, because for example the two Seyfert 1's could be paired each with a z = .28 or .33 quasar across NGC1398. I have taken the interpretation that the two quasars of z = .28 and .33 have been ejected from NGC1398 as in the case of NGC2639 shown in Figure 2-5. Also the z = .04 Seyfert could have been ejected from NGC1398, but is more likely to originate from NGC1385 which is only 20 arc min NW of the Seyfert and about the second brightest low redshift galaxy in the field. But these are details compared to the main conclusion that *this is a cluster of objects at the same distance, but of very different, hierarchically stepped redshifts.* This along with the many other observed configurations which have been discussed so far would seem to me to be utterly decisive.

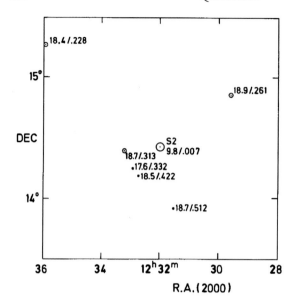

Fig. 8-10b. The brightest Seyfert galaxy in the pictured region is NGC4501 (M88). All quasar/AGN's with .100 < z < .92 are plotted. The z = .261 and .332 are paired across the Seyfert 2 and the rest aligned across the z = .332 quasar. Average out the ejection velocities in order to get close to the magic numbers.

As entertainment for the reader, I include Figure 8-10b. The central galaxy is the very bright Sbc, NGC501 (M88) which it turned out in the end was a Seyfert 2. (Not in the X-ray sample of Chapter 2). Notice the AGN/QSO's of z = .261 and .332 paired across the Seyfert. Then notice the four AGN/QSO's paired across the z = .332 quasar. It is easy to compute the ejection velocities corresponding to .04z and .14z and the intrinsic redshifts which result at around z = .37.

At last, a cluster of Quasars!

As this book was being edited for publication, Geoff Burbidge in his inimitable telephone style, mentioned to me that there was an NGC galaxy next to a bright, famous radio quasar called 3C345. I was excited to find that the galaxy was a Seyfert and immediately looked for the disposition of known quasars around this pair. By great good luck they turned out to be in one of two, sample 8 sq. deg. fields in the sky which had been most thoroughly searched for quasars. David Crampton and collaborators had taken many slitless spectrum plates with the Canada-France-Hawaii telescope on Mauna Kea and identified all the candidates in the field pictured in Figure 8-11.

Now the arrow identifying 3C345 points also to an obvious grouping of quasar candidates. Since these researchers had been actually looking for clusters and associations of quasars, for which they had found some evidence in medium redshift quasars, it is puzzling why they had not investigated the grouping around 3C345. Of course the first question a person would naturally ask is: "Is there anything different about these quasars from those in the rest of the field?" The answer springs out of casual inspection of the catalogued quasars. The ones around 3C345 are brighter and lower redshift than the average in the rest in the field. This is shown in Figure 8-12 where it is seen that for quasars with .5 < z < 1.6, there are practically no quasars in an

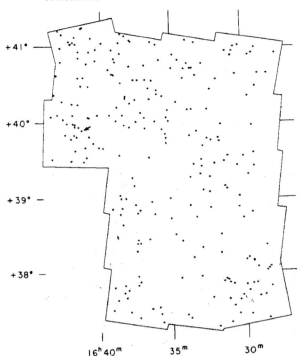

Fig. 8-11. Quasar candidates discovered in a 8 sq. deg. area by Crampton *et al.* (*Astrophysical Journal.* 96,816,1988). The concentration around 3C345 can be seen (arrow). This region is shown enlarged in Fig. 8-12.

equal area of field to the west of the 3C345 group—an area which was searched in exactly the same way!

Because 3C345 is bright and variable it was studied in X-rays with the ROSAT satellite and a number of observations were in the archives. That enables a test related to Chapter 2, where pairs of X-ray quasars were found across active Seyfert and Seyfert type objects. Figure 8-13 shows that the brightest X-ray sources next to 3C345 pair across it, with one of C = 37 counts/ks and one with C = 62 counts/ks. *This is just like NGC4258, NGC2639, NGC4235 and all the other pairs which testify so clearly to the ejection origin of quasars.* But in this case, there are three additional quasars closely in this same

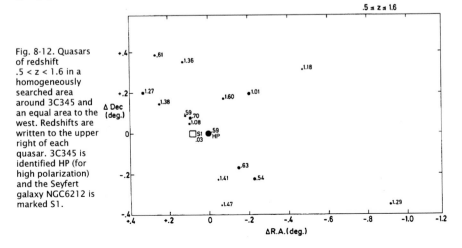

Fig. 8-12. Quasars of redshift .5 < z < 1.6 in a homogeneously searched area around 3C345 and an equal area to the west. Redshifts are written to the upper right of each quasar. 3C345 is identified HP (for high polarization) and the Seyfert galaxy NGC6212 is marked S1.

X-RAY IDENTIFICATIONS

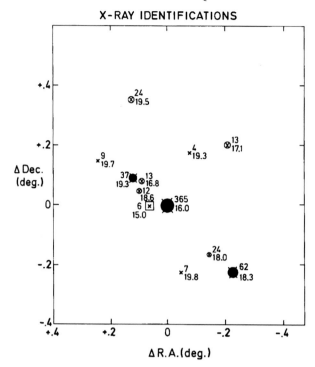

Fig. 8-13. Those quasars which are detected as X-ray sources in archived ROSAT observations are marked with counts/ks written to the upper right. Apparent V magnitude is written below. 3C345 is the brightest quasar in the center of the line of X-ray quasars.

line, giving a compound probability for chance of only 3×10^{-8} (or three in one hundred million)!

Drawing from the examples indicated in the first two chapters, it is simplest to interpret the Seyfert (NGC6212) as the oldest galaxy in the cluster (at z = .03). Then 3C345 at z = .59 (like 3C232 at z = .53 and 3C275.1 at z = .57 from Chapter 1), has been ejected from its nearby active galaxy. In turn, like the earlier examples, 3C345 has ejected a string of X-ray quasars in opposite directions. Notice also how the redshifts which comprise the 14 quasars making up this cluster have redshifts close to three of the quantized values: z = .60, .91 and 1.41. The quasar pairs z = 1.38 and 1.41 and 1.36 and 1.47 are most likely ejected from the Seyfert NGC6212. From the mean deviation from the quantized values I would conclude the two NE quasars of these four quasars were ejected slightly toward us and the SW quasars slightly away from us as we look at NGC6212.

The breaking up of a strong X-ray source into multiple quasars we have encountered before, while analyzing X-ray pairs across the Seyferts discussed in Chapter 2. The most active object here, 3C345, is highly polarized, a strong radio and X-ray source and violently variable—all characteristics of a BL Lac object. So we have *another* case of a bright, BL Lac type object significantly associated with a Seyfert as in Chapter 2. I would suggest that it is in the process of notching down to the next intrinsic redshift of z = .30. The quasars ejected from it will remain at higher redshifts for a somewhat longer time before they notch down to the next permitted redshift. As the cluster does this it divides into numerous pieces which become then the galaxies in

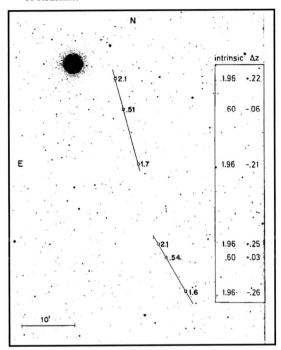

Fig. 8-14. The Arp/Hazard triplets are pictured with their measured redshifts written to the right of each quasar. In the box to the right are written the nearest intrinsic redshift peaks and the velocity components in z which are required to give equal and opposite ejections.

an X-ray Abell cluster as discussed in Chapter 6. As the higher redshifts decay, we wind up with a populous cluster where the Seyfert NGC6212 becomes the older, larger, slightly lower redshift, central galaxy. *It is proposed that we are here seeing the evolution of a group of quasars into a cluster of normal galaxies.*

I also have the feeling that when the object is about to notch down another step in redshift that it goes into the stage of breaking up or fissioning. In fact, if its intrinsic redshift is primarily a function of its age, and hence mass of its fundamental particles, when it notches down in redshift, it must mean that the masses of its particles take a sudden step upward. That would certainly seem reason for a sudden intense rise in radiation, violent variability and ejection or breaking up.

The Arp/Hazard groups of Quasars

About 18 years ago, Cyril Hazard* was identifying quasars on 6 × 6 deg. objective prism plates taken with the Schmidt telescope in Australia. On one plate there were two configurations of quasars which, after I had determined redshifts with the 200-inch at Palomar, showed unmistakable physical associations of quasars of much different redshifts. One was a group of five quasars with redshifts from z = .86 to z = 2.12 (see *Quasars, Redshifts and Controversies* p 64). Another came to be called the Arp/Hazard triplets and consisted of a bright quasar of z = .51 with fainter quasars 2.15 and 1.72 aligned exactly across it and—right next to this triplet—another triplet consisting of a

---

*   This talented radio astronomer was the first to identify the optical counterpart to the radio quasar 3C273 which later turned out to be the brightest apparent magnitude quasar in the sky. He had a difficult career compared to the theorists who misinterpreted this fundamental discovery.

bright quasar of z = .54 with fainter quasars of z = 2.12 and z = 1.61 aligned across it. (See Figure 8-14 here and *Astrophysical Journal.* 240,726,1980.) *Regardless of what these quasars were or what caused their redshifts this proved unequivocally that their redshifts were not measures of their distances.*

And yet extragalactic astronomy has gone on ignoring the evidence and investing more and more money, careers and societal trust in a fundamental assumption which is completely disproven by just a glance at a few published pictures. When pondering Figure 8-14, the repetition reminds us of Chapter 6 where we remarked about the great galaxy clusters, Virgo and Fornax, that nature must know astronomers are not very quick because she shows them everything twice. But if they still do not catch on, it will just have to wait for entities with better judgment.

But now after seeing all the pairs of quasars ejected from Seyferts, and particularly the quasars ejected from 3C345 at z = .59, a beautifully clear understanding of what is happening in the two triplets in Figure 8-14 is possible. The central quasars in the triplets (at this brightness and redshift probably, more like BL Lac's or compact young Seyferts) are ejecting objects of intrinsic redshift z = 1.96—one in a direction somewhat away from us, and one with a component somewhat toward us. In the upper pair the projected ejection velocity is about .07c and in the lower pair about .09c.** This combination of known intrinsic redshift quantization and now typical ejection velocities explains within a few hundredths the observed redshifts in the two configurations.

But there is a delicious pattern evident in the triplets of Figure 8-14. In both cases the quasar ejected away from us is at a closer projected distance from the central body than the one ejected toward us. Would it not be logical to conclude that because of the light travel time to the further quasar is greater, we are seeing it at an earlier time when the quasar has not traveled as far from the ejecting central object as the quasar that is coming toward us? This works qualitatively very well to explain the sense of the unbalanced distances in Figure 8-14. The measured distance ratios would be in better quantitative agreement, however, if the ejection velocity was higher. But we will see in the next chapter that the Narlikar/Das mechanism for ejecting newly created matter starts out with matter at zero mass emerging with the velocity of light. As the constituent particles in the matter gain mass they slow down in order to conserve momentum. So the quasars have been going faster than the present, decelerated velocity which is now observed. Quantitative calculations are being carried out for this model, and it will be interesting to see how closely they agree with the observed asymmetry of the triplet pairings.

** If a quasar of z = 1.96 is ejected away from the observer with a projected radial velocity of v = .064c then the observed redshift will be $(1 + z_i)(1 + z_v) = (1 + 1.96)(1 + .064)$ yields z = 2.15 observed for the first quasar in the triplet, and so on. In order to make the forward and away velocities balance one should assume the central quasars are coming slightly toward us which would make the intrinsic redshifts of the central quasars z = .53 and .62, closer to the quantized value of z = .60

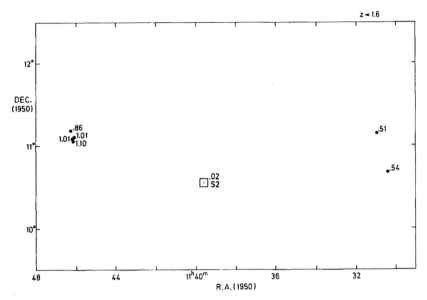

Fig. 8-15. The area in which the Arp/Hazard group and triplets are found. Only redshifts less than z = 1.6 are plotted. The central box identifies one of the 13 most luminous X-ray Seyferts known over the whole sky. (If it were at its redshift distance.)

It is also tempting to consider that this slowing down process continues as the quasars evolve into galaxies, and that by the time they approach the stage of what we call normal galaxies, that they are locked into a primary rest frame in what appears to be a very quiet universe as far as peculiar velocities are concerned.

Finally there arises the question: "Where did the two Arp/Hazard groups come from?" We would expect a Seyfert galaxy. So we look in the list of catalogued Seyferts. As Figure 8-15 shows, there is one midway between these two extraordinary groups of quasars. But this is not just an ordinary Seyfert. It is one of the brightest infra-red sources and *one of the 13 most luminous X-ray galaxies for its redshift in the whole sky.* Now that's an active Seyfert!

The Seyfert is NGC3818 and it is situated in a cluster of NGC galaxies. Here we see a low redshift cluster of galaxies, or a galaxy within the cluster, giving birth to new clusters of galaxies. As the new clusters age and approach lower redshifts they should form a string or filament of clusters, as in the X-ray clusters shown in Figure 6-6, and as observed generally for clusters of galaxies on the sky.

A Crucial Test—The Galaxy Cluster Abell 85

Very important results appear in the *European Southern Observatory* (ESO) *Messenger* as progress reports because new observations are reported without being fitted in detail into conventional theory. One such result was reported by a group of French observers in 1997 (No. 84, p. 20). It presented redshift measures of a large number of galaxies in the X-ray cluster of galaxies, Abell 85.

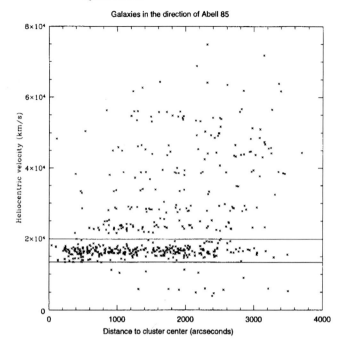

Galaxies in the direction of Abell 85

Fig. 8-16. The X-ray
galaxy cluster Abell
85. Individual galaxies
are plotted as a
function of their
redshift and distance
from the cluster
center. Measures by F.
Durret, P. Felenbok, D.
Gerbal, J. Guibert, C.
Lobo and E. Slezak.

The first thing we notice about this cluster is that its listed redshift is z = .055.
One could not get much closer to the first quantized quasar redshift peak of z = .061.
But then Abell 85 is a very strong X-ray cluster and the X-ray clusters are extraordinarily
sharply peaked at z = .06 (see Figure 6-17). The second thing is obvious from the
distribution of the redshifts of its galaxies as shown in Figure 8-16. *The redshifts are
discretized!* The observers remark that this "could correspond to voids and sheets of
galaxies and could therefore be used as an indicator of large scale structures in this
direction." *But a glance at the Figure shows that the groups of galaxies at higher redshifts are
concentrated more to the center of the cluster than background galaxies.*

It is true that as the redshifts of the galaxies get higher, that they spread to further
distances from the cluster, but never as far as the highest and lowest redshifts measured
which are the best determination of background distribution. For example, the
observations register nicely the sheet of galaxies that pervades half of the sky at
cz = 5000 to 6000 km/sec (the Perseus-Pisces filament phenomenon discussed in
Chapter 6). But that particular value of preferred redshift, lying below the cluster
galaxies in Figure 8-16, is distributed conspicuously further away from the cluster than
the first two redshift peaks lying above the cluster at cz = 23,000 and 28,000 km/sec.
This again is incontrovertible evidence, that can be seen by just looking at a picture, that
intrinsically higher redshifts belong to this cluster and that they are quantized.

If clusters of quasars evolve into clusters of galaxies, the crucial test of this process
would be to see the quantization of quasar redshifts be reflected in the quantization of
galaxy redshifts. And there it is—the quantized redshifts in Abell 85! What else could

explain the redshift quantization in the cluster, and in fact, in low redshift galaxies in general?

Of course as the large redshifts decay, the quantization steps must come closer together. But the fact that the quantization is present in the same kind of an X-ray cluster that was associated with low redshift, active galaxies as discussed in Chapter 6, supports the conclusion that these clusters are intrinsically redshifted objects like the quasars. These clusters would seem to represent the natural next evolutionary step from the groups and clusters of quasars which are associated with the same parent galaxies.

Redshift Quantization as a Function of Electron Spin

The factor of redshift quantization for the quasars is 1.23 as in:

$$(1 + z_n)/ (1 + z_o) = (1.23)^n$$

which gives:

$$z_1 = .06 \quad z_3 = .60 \quad z_5 = 1.41 \quad z_7 = 2.64$$
$$z_2 = .30 \quad z_4 = .96 \quad z_6 = 1.96 \quad z_8 = (3.47)$$

These peaks are observationally so well established that it has always been a great frustration for me not to be able to use the factor 1.23 in the 72 km/sec quantization which is observed for the galaxies, namely:

$$z = 72, 144, 216, 288 \ etc.$$

For example $(1 + z_1)/1.23$ yields a large negative redshift, not 72 km/sec = .00024. So just out of curiosity I calculated what the power of 1.23 should be in order to give a redshift of 72 km/sec :

$$1 + 72/c = 1.00024 = (1.23)^a$$

It turned out that the value of a = .0011592. (Details of this development can be found in *Apeiron*, vol.2, no. 2, p.43, April 95.)

Because I had been exploring the spin of the electron as a possible basic time unit, I was in a position to notice the extraordinary coincidence of this power, a, with the numbers in the value measured for the magnetic moment of the electron (which is ½ the Landé g splitting factor):

$$g/2 = 1.00115965 = 1 + \alpha/2\pi$$
$$a = .001164[4]$$
$$\alpha/2\pi = .00116141$$

where a is now the power to which 1.229 must be raised (1.229 is the most accurately the quasar redshift factor can be measured). $\alpha/2\pi$ is the fine structure constant that determines the line spacing in atomic spectra. Considering the difficulty of picking five correct numbers in a row (like a lottery)—there seems to be something significant here. But so far it is not a solution, but only a clue which connects the quasar redshift spacing with atoms in quantized states.

In an attempt to make some sense of this I tried to visualize an electron with its spin interacting with the magnetic field of the nucleus of its atom. Depending on its spin orientation, it can assume a series of quantized, fine structure energy levels. At an earlier time the electron wants to be at a lower mass (because of m varying as t^2). But its least change is the lowest permitted quantum step so that when it does notch lower, it forms an electron which is less massive by $(1.229)^{-.001164}$ than at our epoch. Then any atomic transition, emitting or absorbing a line, will be redshifted by 72 km/sec relative to our terrestrial standards. The second notch down will give +144 km/sec, then 216 *etc.* as observed in our Local Group companion galaxies which are of the order of 10^7 years younger than our parent galaxy M31.

Taking into account the interaction of the elementary particle with the surrounding electromagnetic field is the domain of quantum electrodynamics and the language becomes very specialized. Phenomenologically, we can say that the increase of the intrinsic redshift in quasars as we consider younger matter seems to come from the lower particle (electron) masses. In the other direction, as the matter ages, the quasars show that their electrons do not increase in mass smoothly but rather in quantized steps of a factor of 1.23. The evidence from the smaller intrinsic redshift steps indicates that there is fine structure between these large steps—that the redshift drops in smaller quantized steps, but probably spending the most time in the strong resonances of the 1.23 factors. But what the factor 1.23 represents, or where it comes from, is very difficult to say at the moment. Perhaps we need more empirical clues.

Mass as a Frequency

In order to attempt to understand why masses should be quantized one has to ask the question of what, fundamentally, mass is. The evidence discussed in this book seems to indicate that the elementary masses change with time. The interesting question then becomes what is the operational definition of time? One suggestion would be that time is measured by the regular repetition of a configuration, like the rotation of the earth or its revolution around the sun. The most fundamental definition of time might then be the rotation, or spin, of the electron. (For the present purpose it does not seem to matter whether the electron is some unspecified distribution of charge spinning about an internal axis or a loop of circulating current as some have modeled it.)

Can the mass of the electron then be expressed in units of time? Formally this is simply achieved by using the Compton frequency of the electron, ν_C, the Planck constant, h, and the velocity of light, c.

$$m_e = h/c^2\, \nu_C$$

The terrestrial Compton frequency is 1.2356×10^{20}/sec (apparently no connection to the 1.23 quasar redshift factor). So we see that as the mass grows, the frequency (the fundamental clock rate) increases. In the past the frequency was lower. Looking out at younger galaxies we are seeing them in an era when their clocks were running slower. The redshifted photons are just carrying information on the clock rates where they came from. Particle mass then might be considered as the fundamental frequency of matter created at a particular epoch.

This way of describing the observations seems to have another advantage. Namely, the frequencies of spinning electrons are known to be quantized. Therefore if they are to change they would have to change in discrete steps. Time itself would, in a way, be quantized. Perhaps the quantum mechanical properties of materializing and dematerializing (virtual) matter would be related to the steps in time. But as for the various values observed for quantized redshifts (considered as frequencies) the possibilities of beat frequencies, fundamentals and overtones would seem to present rich possibilities for explaining the patterns observed.

Mass Quantization in Quasars, Planets and Particles

In 1990 an artist friend in Tenerife, Jess Artem, mentioned to me that the Titius-Bode law expressing planetary distances from the sun obeyed quite well a series based on the preferred quasar redshifts. The so-called Bode's law has been so much discussed and criticized that I was at first skeptical. But then I noticed in a book by T. F. Lee that the ratio of the mass of earth to Venus was 1.23 and that he claimed that powers of this factor gave a limited Bode's law. Looking back now I realize that he had no knowledge of the 1.23 factor in quasar redshifts.

Checking into this for myself, I used the most modern compilation of planetary masses and found the amazing result that the ratio of the masses of all nine planets fell very close to integer powers of the factor 1.23. The first thing one thinks is: "Could this be an artifact of the calculation?" But by varying measured mass values uniformly one could check that they would spread evenly between n and the next nearest integer. A simple test showed the observed distribution had *less* chance than one in 1300 of being accidental. (The complete analysis can be consulted in *Apeiron*, Apr. 95, p.42).

Just to push the relation to a ridiculous extreme, I calculated the mass ratio with the sun. It fell within 9% of an integer value when 50 % was expected by random. The mass ratios of the satellites of Earth, Jupiter, Saturn and Uranus fell even closer (interestingly, nine out of eleven on half integer values). The obvious implication is that since this same mass ratio applies to the planets, satellites and sun, and also the electrons in the quasars, that masses on all scales, at least in our local universe, are formed in the same ratio. This would suggest a rather audacious test: namely, are terrestrial electrons in a ratio of $(1.23)^n$ to the mass of the earth? *That ratio turns out to be within 6%.*

Quantization of Planetary Orbits

As for Bode's law of planetary distances (using terms that varied as 2^n), it failed badly with the discovery of Neptune and Pluto. It was modified into the Blagg-Richardson law involving $(1.7275)^n$ with complicated corrections for each planet. From the quasars the mass/redshift factor is estimated to be determined as $1.2288 \pm .0006$. The most accurate value I determined from the fit to the planets, satellites and terrestrial electrons is 1.2282. Table 8-2 here shows the application of the factor (1.228) to the planetary distances. The fit to integer values has only one chance in 500 of being accidental. The fit to the four planets next outward from the earth is particularly good. On the other hand the fit to the Blagg-Richardson mean law is just what one expects from chance. This is useful in both showing that the modified Bode's law is not meaningful and also in showing that a fit with an arbitrary factor like 1.7275 is no better than that expected from a random distribution of numbers, unlike the fit achieved with the factor 1.228.

There are, however, some less than satisfactory features of these numerical fits. For example, there are large numbers of missing integers, *i.e.* integers where there is no body present. But overall I feel there must be some significance to it. Probably a more correct formula needs to be derived. But as it stands it is the kind of result that drives most scientists crazy. They glower around that anyone who talks about this is a "numerologist." A very derogatory term for a scientist. Probably even worse than getting the "wrong" answer for many scientists is to live with uncertainty.

Recently, however, exciting results have been brought to my attention. One is a fit to the mean planetary radii by using an angular momentum (Bohr) type condition. Saulo Carneiro of the Universidade de São Paulo has presented calculations by Oliveira Neto of the Universidade de Brasilia showing that the principal quantum numbers squared from n = 1 to 10 represent very well the orbital radii of the planets and asteroids from Mercury to Pluto. Venus and Earth are then represented by additional orbital quantum numbers 0 and 1. Two Italian physicists, A.G. Agnese and R. Festa, and a French astronomer, L. Nottale, get a very similar planetary system with orbits like a Bohr atom which is quantized as n^2 or $n^2 + 1/2n$. Figure 8-17 shows the excellence of the n^2 fit to the known orbital radii of the planets. Both solutions use a gravitational Planck quantum of action (or fine structure constant) scaled up from the electronic fields governing atoms to the size of a planetary system.

A. and J. Rubcic, University of Zagreb, have also presented a very good fit to planetary orbits with a formula $r = r_1 n^2$ where n is a consecutive integer number. The fit depends on the specific angular momentum of each planet being quantized. The long felt analogy of the solar system to an atom has now received some quantitative support from scaled physical laws. This could represent a profound step in physical

Table 8-2 Orbital Sizes in Solar System

Planetary Distance (semi Major Axis in AU)		factor 1.228		factor 1.7275	
		n	$\|\varepsilon\|$	n	$\|\varepsilon\|$
Mercury	.387	−4.62	(.12)	−1.74	(.24)
Venus	.723	−1.58	(.08)	−.59	(.09)
Earth	1				
Mars	1.524	2.05	.05	.77	.23
Asteroids	2.8	5.01	.01	1.88	.12
Jupiter	5.203	8.03	.03	3.02	.02
Saturn	9.539	10.98	.02	4.13	.13
Uranus	19.191	14.38	(.12)	5.40	(.10)
Neptune	30.061	16.57	(.07)	6.23	.23
Pluto	39.529	17.90	.10	6.73	(.23)

ε values in parentheses are deviations from half integer values.

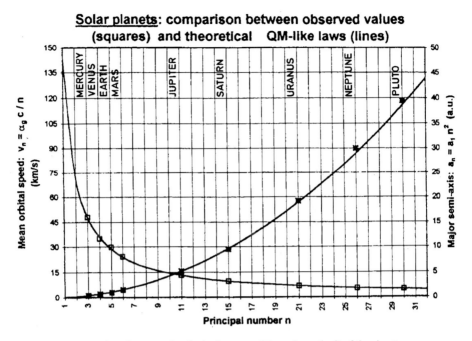

Fig. 8-17. This adaptation of a plot by Agnese and Festa shows the fit of the planetary distances from the sun to the n^2 quantization law. On the same graph is shown the fit to the velocities of the planets in their orbits to 144 km/sec divided by n.

understanding of nature.

I have two comments about these developments: One is that if the series of numbers n^2 is correct, then the planets are not situated at random. Other series such as Bode's law (2^n) or the fit with $(1.23)^n$ are also ordered and could give significant fits for certain ranges. But the most meaningful law would be the one that fit all the data for consecutive numbers staring with n = 1, 2, 3 *etc.* It would be interesting to explore the mathematical relation between the n^2 law and Bode's law to see how close he, and others, came to the right answer. It clearly would have been more rewarding to find the correct expression than to scorn the early researchers as "numerologists."

The second point is that for the solution to work one must use a *mean* mass for each planet. What this signifies to me is that originally the planets must have been of the same seed mass and then accreted from particles in the same orbit or grew by some other means to their present differing sizes. Each accreting particle would add its angular momentum, and so Jupiter which has most of the solar system angular momentum would be made up of the most elementary particles. If there were some small, uniform, potential particles in the beginning, then one could imagine them being in quantum mechanically governed orbits such as in an atom. But, of course, this would have fascinating implications about the very most initial birth of the solar system.

Formation of Discrete Bodies in the Universe

The assumption in cosmogony until now has been that all bodies in the universe condensed out of a uniformly spread, homogeneous medium and hierarchically aggregated to their current sizes. The evidence, however, is that proto-bodies are ejected from previously existing parent bodies and subsequently grow to their presently observed sizes. We have seen this strongly in the formation of galaxies, quasars and clusters of galaxies and quasars (for example the ejected knots in M87 as shown in Plate 8-18). We are seeing it now in the formation of planets. Pictures of T Tauri stars, acknowledged by astronomers to represent stars in the process of formation, show extremely thin jets with condensations along their length and at their ends (Plate 8-19). Even the formation of elementary particles seems to follow this rule as inferred from the break up of initial, relatively massive Planck particles into electrons, positrons, *etc.*

Throughout these processes we see production of bodies of discrete properties— *i.e.* quantization. Though the rules of the relationship of their scales may still be a mystery, the evidence of the quantization of the planets of the solar system appears to be a demonstration that planetary systems do not form from the collapse of a solar nebula. There is no apparent way to obtain ordered discreteness from a formless, diffuse cloud. So the evidence for quantization in solar system planets seems to be another contradiction of the conventional assumption and evidence for the emergence of material from a previous, parent body.

The Problem of Quantization and Velocities

Another astonishing relation was pointed out by L. Nottale, namely that the velocities in the planetary orbits (at least out to Uranus) decrease as 144 km/sec divided by 3, 4, 5, 6, 11, 15, 21, 26, and 30. Agnese and Festa obtain the same exact fit all the way out to Pluto as shown in their plot reproduced here in Figure 8-17.

Amazingly, as we saw in the beginning of this chapter, 144 km/sec is a prominent quantization number in the redshifts of galaxies! But the mind blowing aspect of this is that there are a number of arguments why the 144 km/sec redshift peak in galaxies cannot represent a velocity at all (for example, random orientations of the velocities would smear it out). If I had to guess, I might think that particle masses change in discrete steps which means that fundamental scale lengths change in steps. If scale lengths in a primeval planetary system change in steps, then Kepler's third law would require periods (velocities in orbits) to change in steps. The challenge would be to quantitatively evolve the particle physics laws to gravitational physics laws as a function of time.

The evidence points to quasars being ejected with initially high redshifts and high velocities and by the time they evolve to redshifts around z = .6 they have slowed to velocities of about .1c. The evidence also indicates that they continue to evolve into normal galaxies. But galaxies have much smaller velocities than .1c and the quantization of galaxy redshifts into smaller periods such as 37.5, 72.4, 144, 216 km/sec *etc.* requires their peculiar velocities to be less than about 20 km/sec. There seem to be only two possibilities:

1) The galaxies continue to slow their velocities as they evolve to lower redshifts—*i.e.* they become locked into some large scale rest frame structure, or

2) Some wave interference prevents us from seeing the galaxies when they are not at the observed velocity peaks relative to us (as in the machion proposal earlier in this chapter).

In either case we have a very new and a very different view of the universe.

General Laws on all Scales?

As a vivid demonstration that the phenomenon of ejection of discretized bodies not only characterizes the birth of quasars and galaxies (as shown by the protogalaxies emerging from M87 in Plate 8-18) we show here in Plate 8-19 and Plate 8-20 that it also extends down into the realm of young star formation. If we rotate the directions of the jets to be parallel, there is an uncanny resemblance between the formation of young galaxies and the formation of young stars. Note the opposite ejection, its extreme collimation, the one-sideness of the optical jets, and the discrete, compact objects coming out in the narrow cones from both the active galaxies and the young stars. Does the same mechanism evident in the formation of these young stars extend to the formation of planets?

In the phenomenon of quantization, we have a connection from the redshifts of the quasars, to the redshifts of the galaxies, to the properties of the solar system and finally to the properties of fundamental particles like the electrons. The quantization of physical parameters would seem to be governed by the laws of non-local physics, *i.e.* like quantum mechanics in which the fundamental parameter appears to be time—for example the repetition rate of a spinning electron. It is clear that we are not running out of problems to solve. In fact, contrary to some rumors that we are reaching an end to physics, the more we learn the more primitive our previous understanding appears, and the more challenging the problems become.

Chapter 9

COSMOLOGY

I f redshifts are not caused by velocity of recession, what are they? The answer to this question goes back to the roots of modern cosmology and opens up the possibility of a whole new understanding of the universe.

Einstein's General Relativity

Like most people, I grew up with the received wisdom that Einstein's General Relativity was so profound and complicated that only a very few people in the world understood it. But eventually it dawned on me that the essential idea was very simple, and it was only the elaborations that were complicated. The simplest mathematical expression of G.R. is shown below:

$$(1) \qquad\qquad G_{\mu\nu} = T_{\mu\nu}$$

The T represents the energy and momentum of a system of particles. In order to describe their behavior in great generality, they are considered to be in a space whose geometrical properties (*e.g.* curvature of space-time) are described by G. Now the solution to this equation tells us how these particles behave with time. The important feature of this solution is very simple to visualize, either the initial energy is large and the ensemble continues to expand or the energy is small and the ensemble collapses under the force of gravity. This is the unstable universe which distressed Einstein and caused him to introduce the cosmological constant (a special energy term) which just balanced the universe.

But in 1922 the Russian Mathematician, Alexander Friedmann, put forth a solution in which the spatial separations of the particles expanded with time. At first reluctant, Einstein later embraced the expanding universe solution so enthusiastically that he renounced his cosmological "fudge factor" as "the greatest blunder of my life." The Lundmark-Hubble relation was in the air at the time, and it seemed an ideal synthesis to interpret the redshifts of the extragalactic nebulae as the recession velocity of their expanding space-time reference frame. But basically, the theory was that the galaxies at our time were expanding away from each other, and therefore must have all originated in a "Big Bang"—that is, the universe was created instantaneously out of nothing.

Simple folk would say there is no free lunch. Philosophers would argue that nothing is nothing, and it does not become something. But the error in the science was not found until 1977 (and of course most scientists still vehemently refuse to admit it was an error). I believe the error was in the assumption that the particle masses remain constant in time. No one ever saw an atom or electron growing heavier with time. So it was natural for humans to assume their tiny little slice of space and time was the way the whole universe was.

General Solutions

The reason I believe that particle masses change on cosmic time scales is the following: First, Jayant Narlikar showed in a crucial step in 1977 (*Annals of Physics*, 107,325) that if (1) is written in a more general form, it would contain terms involving particle masses, m, which were not constant over all spatial distances and intervals of time. In this case a solution of the more *general* relativistic equation is:

(2) $m = at^2$,

i.e. the mass of an elementary particle varied as the time squared (where a is a constant).

Now mathematicians teach that the proper way to solve an equation is to solve in general terms *before* any approximations are made. After the general solution is made, approximations like m = constant can be made *if suited to the problem*, for example, terrestrial problems which involve relatively short time intervals. The result of the approximation which Friedmann made in 1922 in order to solve the G.R. equations was to force the observed effects of the actually changing mass into the geometrical terms on the left hand side of the equation. That leads to many models of curved space-time. But if the changing mass is explicitly expressed, there is no need for curved space-time. (Also see Appendix C).

In the Narlikar solution, the geometrical terms on the left side of the equation lead to "flat" space time. That is, the coordinates of a particle are simply measured in three orthogonal directions and there is no physics in this operation. There may be computational advantages in the presence of strong gravitational fields to transform to non-Euclidean geometries, but in the weak gravitational fields of the cosmos it is reassuring that the dynamical solutions are simple and straightforward.

In Appendix A, a schematic mathematical derivation of this solution to the field equations is indicated with some comments on what I think are the key points.

Redshifts as a Function of Time

The most useful feature of the Narlikar solution is that it explains the preceding book full of observations. If particle masses are a function of time, then younger (more recently created) electrons have smaller masses. When a less massive electron makes a transition between atomic orbits, the photon involved has lower energy and the resulting spectral line is redshifted. The consistent lesson of the observations we have discussed is the younger the object, the higher the intrinsic redshift.

Actually this empirical, observational result enables us to derive the whole general solution without recourse to any relativity or tensor calculus at all, but only the simplest of calculations. One might even call this the philosophical derivation. It starts by considering a single electron at the moment of creation. It has no mass because it cannot be compared with anything else. When it communicates with another electron it acquires the property of mass. As it ages the light signal sphere within which it communicates enlarges. It thereby communicates with more and more particles and acquires more and more mass. The light sphere enlarges as r^3, the interaction weakens as the potential, $1/r$, so the mass increases as $r^2 = c^2 t^2$. (Appendix B presents a schematic derivation from simple differential and integral calculus). At the end of this chapter we mention briefly what effect a clumpy, rather than a uniform-density universe might have.

With a simple logical model we have obtained the same result as the general solution of the general relativistic equations. But this general solution is much, much more powerful than the conventional solutions because it is Machian.

Machian Physics

At the Cal Tech luncheon table I once asked the reigning expert to define Mach's Principle. After thinking all lunch he surprised me at the end by saying it was too difficult to explain. I will try to do it anyway. I think it is generally considered that Mach maintained that matter at great distances in the universe from us influences our local physics. The oft quoted example is that when the subway stops with a jerk, it is the distant stars that throw you down (your inertia is a result of communication with those distant bodies).

The importance of this for the discussion here springs from the conventional treatment of the general relativistic equations. Einstein himself started with the conviction that Ernst Mach was correct. But at the end of the day he had to sadly admit that his equations were not Machian and that general relativity was a "local" theory. But we have seen that the equations were not wrong (after all they just represent conservation of mass-energy and momentum). It was the fact that particles realized their mass by communication within their creation light sphere that made the physics Machian—and that had been omitted in the conventional solution.

This becomes terribly important from another aspect, namely quantum mechanics. In the small mass-energy regime, discrete rather than continuous phenomena are encountered. Empirically this is a well-validated physics. But to the despair of generations of physicists, it appears impossible to unify general relativity and quantum mechanics. Perhaps the outstanding aspect of quantum phenomena, however, is that they involve non-local physics. If we make classical dynamics a non-local theory then we open the prospects of unifying these two branches of physics.

Creation of Matter

Another long-standing but little emphasized embarrassment of the conventional relativistic treatment was the existence of singularities, especially in space-time regions where $m = 0$. Singularity is a euphemism for "the physics just breaks down." It was a particular strength of the Machian solution then when the Indian Astrophysicist Kembahvi showed the singularities turned into the zero mass hyper surfaces in the variable mass formulation. What had been a drawback for the older approach then became a necessity for mass creation.

Of course if the universe is operationally defined as everything that is detectable or potentially detectable there can be no such thing as "new" matter. So when we speak of creation of matter we do not mean matter coming into our universe from somewhere else (there is nowhere else) or from nothing. We must mean the transformation of previously existing mass-energy. Probably this means materialization from a previously diffused state—a concept which would relate well to quantum physics.

We know from the X-ray observations of Seyferts that the quasars, explainable as younger matter, emerge from the small dense nuclei of active galaxies. That is obviously the place where the new matter is created or materializes. This vindicates the reasoning of the famous physicist Paul Dirac who considered two types of matter creation; one in empty space and one in the presence of preexisting matter. This was daring for his time. In the last few years Jayant Narlikar has been exploring matter creation with conventional relativistic physics near mass concentrations.

The important point seems to be to distinguish between the broad outlines of the models involving matter creation and the conventional models involving black holes and accretion disks. The greatly publicized theory is black holes where everything falls in. But the observations show everything falling out! (Can we count on conventional science *always* choosing the incorrect alternative between two possibilities? I would vote yes, because the important problems usually require a change in paradigm which is forbidden to conventional science.)

Accretion Disks and Black Holes

Accretion disks seem to be a natural occurrence in stars where the burning of the fuel produces a dense core with high gravitational attraction on surrounding material. The material appears to spiral into the star in a plane called the accretion disk. A great deal of analysis in astronomy has gone into interpreting the outbursts observed in violently variable stars as the burning of incoming lumps of material as they hit the hot accretion disk. The match with observations is fairly successful.

But it was almost immediately speculated that the material spirals in to form an even denser object, which was hypothesized to be a black hole. The salient property of this theoretical beast is that anything that falls into it can never come out. There are several things that can be said about a black hole. The first is that when you let r go to zero in Newton's famous force equation $F = GmM/r^2$, you get infinity—in other words a singularity where physics, as we experimentally know it, just becomes meaningless. In the very complex equations which have been developed to handle such difficulties, one amazing result that always brings appreciative "oohs" from the audience is that if you watched someone fall into a black hole that it would take an infinity of time for him to disappear. How then, in the accepted age of our universe, can you form a black hole? I asked this of a good physicist friend at lunch once and at the end he said, "Well you may never get a black hole but you can come as close to it as you want."

But it is important if accreted material can never get "inside" a black hole. That is because we have to eject material out of the supposed black hole accretion disks in the centers of active galaxies.

Ejection of New Matter

In an accretion disk around the nucleus of an active galaxy it has been proposed that lumps of matter crash into the disk. That might account for some of the smaller light variations observed in active galactic nuclei, but how do you get the long, collimated ejections in opposite directions that are observed? If a cloud of material falls onto an accretion disk it goes splat. At best the material follows the lines of magnetic force anchored in the disk and goes in every direction but up and down.

But even if you could get infalling material actually inside a black hole, how in the world would you get it out? The material has mass and nothing can fall out of a black hole. What we need is a white hole—a place where everything falls out of. Fred Hoyle has always said that mathematically, a white hole is just the time reversal of a black hole. Originally Hoyle used the term negative energy field and recently Narlikar has used the concept to describe a situation where the mass concentration in an active galaxy nucleus becomes so strong that it triggers energy inflow from the extended negative energy field in which it is imbedded.

This model has the advantage that it naturally leads to an influx of energy which expands or blows apart the mass concentration thus stopping the inflow of new material. Subsequently the core can recontract and the process can recommence. This is an ideal explanation for the observed intermittent ejections of material from the nuclei of galaxies. In fact, the quantization of the redshifts observed for the ejected material could well be related to its periodic creation, since the redshift depends directly on the particle masses at the time they emit the photons received by us. The epochs of creation would be imprinted on the particles, which would appear as steps in the intrinsic redshifts. This process is also a potential link to the quantum mechanics that must be an essential part of the creation process.

A further advantage of this white hole scheme is that the new matter is created at the very center of the mass concentration where the spin axis represents the direction of least resistance and can channel it out in opposite directions. This is just the region which is forbidden to access in the conventional black hole. Of course, as the new material passes outward from the nucleus it undoubtedly entrains older material from the parent galaxy especially magnetic fields. These magnetic fields could act as force tubes constraining outflowing ionized gases as they condense into new objects.

The Form of the New Matter

It is challenging to try to imagine what the bulk properties of the newly created material might be. In Chapter 6 we hypothesized that the high-energy radiation from the Local Supercluster center came from the breakup of recently formed Planck particles. The Planck particles started with zero or near zero mass, but the end-product protons and electrons would still have relatively small masses. Since the scale of the particle varies inversely as its mass we would expect a large interaction between low-mass particles. That is we would expect a material with fluid properties.

It is interesting in this respect to recall Viktor Ambarzumian's conclusions in the 1950's from inspecting Schmidt survey pictures of galaxies. He adduced that new galaxies were formed in ejections from older galaxies and suggested that the ejections were initially in the form of a "super fluid." The subsequent radio ejections, observed in classic cases such as Cygnus A shown in here in the Introduction and some of the arcs in clusters shown in Chapter 7, are evocative of how a fluid might behave. In this same vein a compatriot of his, Vorontsov-Velyaminov, concluded from the same survey pictures that galaxies could fission (the opposite of the great modern fad of merging). We noted earlier the evidence for quasars breaking up into X-ray clusters of fainter galaxies. It might be pleasantly instructive to contemplate that, although we may never grasp the ultimate detailed theory, that we are capable of understanding the essence of matters by careful observation and analogy of empirical patterns.

Other comments of potential interest are that the emergence of ionized, low-mass particles along magnetic field lines would wring out the energy of synchrotron emission* extremely efficiently. As we know the young active objects have uniquely high amounts of synchrotron energy. (It is assumed that the ions are created with normal values of electric charge which dominates the spiraling around the magnetic force lines because of the low particle mass.) If the matter is exactly collimated by the field lines it is in a sense "cold." It does not have random velocities of kinetic temperature. This may have something to do with the large amounts of cold molecular gas surprisingly found near the centers of very active galaxies. It is also reminiscent of the cold Big Bang as espoused by David Layzer, which is interesting because the creation we are discussing is in many ways a recurrent "mini bang."

Another matter of interest is that the new matter must initially emerge with the speed of light because, being at zero mass, it is essentially an energy wave and traveling

---

*   due to deceleration of charged particles.

with signal velocity. It will slow down as it gains mass as calculated by Narlikar and Das as discussed later. But while in its low mass state, if it impinges on appreciable material on its way out of the ejecting galaxy, it could, because of its low mass, be trapped inside. If it can coherently condense inside, it is possible that objects in different states of evolution and different redshifts could be blown out together in subsequent expulsions.

Decelerating Ejection

Actually there is a rather complete mathematical theory worked out for what happens to the new matter with zero mass when it is ejected at the speed of light from the center of an active galaxy nucleus. This was given by Jayant Narlikar and P.K. Das in 1980 (*Astrophysical Journal* 240, 401). They show that as the particles in the ejected matter gain mass, they slow down in order to conserve momentum. They decelerate and, depending on the mass of the ejecting nucleus, they either escape or slow to a captured orbit at about 400 kpc from the parent galaxy.

It has been mentioned earlier that by the time the quasar has evolved into a fairly luminous object with an emission line spectrum, that it is in the $z = .3$ to 1.0 range with an ejection velocity of the order of $cz = 30,000$ km/sec. The next stage of evolution seems to be into BL Lac and galaxy cluster objects which have generally lower intrinsic redshifts. *There are many of these kinds of objects actually observed to be associated within about 400 kpc from the parent galaxy as predicted. (See particularly Figure 3-27 and Figure 9-3.)*

The Narlikar-Das calculations apply to the most favorable case for escape from the ejecting galaxy—exit along the minor axis with no interaction with the galaxy and no subsequent interaction with the intergalactic medium. We have argued, however, that the initial low mass state and possible cohesive properties of the ejected objects would lead to increased stopping effects particularly at small angles to the plane of the ejecting galaxy. Therefore a higher percentage of quasars would be captured and at a closer distance. This would give more time for the evolution into hierarchical families of galaxies. These considerations also raise the question of how much residual velocity the new galaxies are eventually left with. The latter is important because the observed quantization of galaxy redshifts requires rather low peculiar or orbital velocities.

Formal Aspects of the Theory

What we have done so far is typically theoretical science—connecting together a number of things that we think we know about, with something we don't know about by giving it a name like "negative energy field." It would be more honest to say "something happens inside the nucleus of an active galaxy causing it to eject material which evolves into new galaxies."

But of course many people have preconceptions and even agnostics have far from unanimous opinions. So any new theory will be challenged by the first test a theory must pass—and a very legitimate test it is indeed—namely the reflexive criticism "The accepted facts disprove your theory." This is where mathematics is very useful. In a short paper (*Astrophysical Journal* 405,51,1993) Narlikar and Arp present the formal

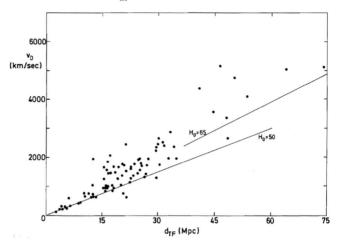

Fig. 9-1. A plot of measured redshift versus distance (the latter from an estimate of the luminosity from the mass of rotating galaxies or distance-Tully-Fisher). The plot shows that low redshift galaxies give a Hubble constant of 50 km/sec/Mpc whereas higher redshift galaxies give larger Hubble constants.

variable particle solution and show how it fits the data better than the Big Bang mantra. I will try to put in words what the equations show in a more economical but specialized language.

Is the Universe Expanding?

Most people would immediately claim that the redshift-apparent magnitude relation for galaxies proved the universe was expanding. But distances to galaxies are large and the light takes an appreciable time to travel to us, so we see the galaxies as they were when the light left them, that is younger. It turns out that for galaxies created at the same time, but seen at different distances, the $m = at^2$ solution requires these galaxies, which are seen at younger stages, to have exactly the same redshift as observed in the Hubble relation. In fact the slope of the relation, *the observed Hubble constant, is predicted within its error of measurement by only one number, the age of our galaxy.* The fact that there is no effect left over to be interpreted as a distance-velocity relation means the universe is not expanding!

This result is supported by all the previous evidence that most of extragalactic redshifts are intrinsic and not velocity. It also eliminates much of the need for the never detected "dark matter." (As described in Chapter 1 after NGC3067 and in Chapter 7 in the discussion of the Coma Cluster.) Moreover the variable mass solution predicts properties of the Hubble relation which the Big Bang cannot account for. For example, Figure 9-1 shows a redshift-distance plot where the distances are derived from the assumption that galaxy rotation velocities are only determined by the mass of the galaxy (the Tully-Fisher relation). Here is a clear result of their assumptions that the expanding universe adherents cannot accept—the Hubble constant increases with distance!

In the variable mass solution, however, among the higher redshift galaxies one finds some born later (*i.e.* younger galaxies). They have increased intrinsic redshifts and stand up above the normal relation. They represent transitions between young galaxies and quasars. Of course, the quasars which violate the Hubble relation so strongly are no longer distant, unprecedentedly high luminosity objects. They are now explained as quite young objects at the distance of the nearby galaxies which they are observationally associated with.

Why Doesn't the Universe Collapse?

The second question people triumphantly put is: "If the universe isn't expanding, why doesn't it collapse?" As Einstein, and Newton before him knew, a static, matter filled universe should fall together. But it turns out that the new factor, the variable particle masses with time, produce mass dependent terms in the dynamical side of the equation which, as Jayant Narlikar has pointed out, guarantee stability. Of course for Big Bangers, who have a universe blowing apart, it is heartbreaking testimony to their plight that they would attempt to make a major criticism of a rival theory because it was unstable!

Local Physics Preserved

Then come a whole host of objections along the lines of "If electrons and protons increase their mass with time, why doesn't the earth spiral into the sun, why don't clock rates increase *etc.* The answer is intriguing and possibly very deep. Mathematically it turns out that the conventional Big Bang solution and our variable mass solution are the same if one makes the conformal transformation

$$(3) \qquad\qquad \tau = t^3/3t_0^2.$$

What this means is that if we operate on the τ time scale, the time on which the matter of our galaxy runs, all the dynamical equations and solutions are the same as in the conventional solutions of the usual relativistic field equations. If we look at another galaxy created more recently, however, their clocks appear to be running slow and their matter appears redshifted.

As time goes on, the particle masses in the different galaxies exchange signals with more and more the same total mass of matter and their clocks asymptotically approach the same rate. This clock time is t, the cosmic time. From the standpoint of the cosmic reference frame, the universe is not expanding. Matter is intermittently materialized into it with clocks that appear to run very slowly at first and then evolve to more normal rates.

From the largest reference frame, the one where the time scale approaches t, the behavior of the sub units, including our own galaxy can be most simply understood. For local matter of our own epoch, all the usual physics operates as we know it—as we measure it on our own τ time scale. Where things go horribly wrong is when we look out from our own galaxy and believe that the redshifts are velocities of recession instead of differing clock rates due to age differences.

The so-called time dilation for objects receding at high velocity is exactly the same function of redshift as for stationary objects whose redshifts are caused by younger creation epochs. This means we expect the same slower decay rates for supernovae light curves as in an expanding universe. A much vaunted proof that the universe is expanding is that the surface brightness of galaxies varies as $(1 + z)^4$—but this too is the same in both theories since the mathematical equations are conformal transforms of each other.

Particle Pair Creation

As for the creation of matter from a zero mass state, it is often objected that pair creation of electrons and positrons from photons in terrestrial laboratories does not produce low-mass electrons. The answer must be that these photons are localized packets of energy and the created electrons and positrons are local entities—not drawn from elsewhere in the universe. In the theory of quantum electrodynamics (QED) which is used in these problems, it is interesting to note that the mass of the electron is not given by the theory but must be specified by experiment in order to introduce a scale length. This implies that a longer scale length for the experiment should set a lower mass for the electron.

As for the vexing problem of renormalizability, the theory encounters infrared divergences as one allows the photon rest mass to approach zero. The cloud of "soft" (long wavelength) photons approaches infinity. Perhaps this longstanding difficulty of infinite electron mass from the theory of quantum electrodynamics has been telling us something important about the connection of electron mass with the universe at large.

Advantages Over the Big Bang

In the continuing creation theory, for redshifts small compared to 1, the Hubble constant is simply $H_0 = 2/t_0 = 2/3\tau_0$. τ_0 is the age of our own galaxy on our time scale. This is determined from the age of the oldest stars in globular clusters as between 13 and 17 billion years and requires a Hubble constant between 39 and 51 km/sec/Mpc. Observed values of the Hubble constant by Allan Sandage in 1988 and 1991 give between 42 and 56 km/sec/Mpc.

The first thing that can be said is that if the variable mass theory is based on incorrect physics, it is extremely unlikely that of all the possible values it could give, it would give the correct value of the observed Hubble constant. The second thing that can be said is that Big Bang assumption leads to a soap opera of conflicting claims about the value of the Hubble constant. This is because the majority of astronomers try to determine the Hubble constant by observing higher redshift objects where the effect of the supposed expansion dominates over the supposed peculiar velocities. But Figure 9-1 shows that they encounter younger objects which give too high a Hubble constant, in the neighborhood of $H_0 = 70$ to 80. (Sandage and Tammann rely more on local objects which consistently give a Hubble constant of about $H_0 = 50$ even though their long distance scale is probably less correct).

The majority of astronomers then, in spite of the fact that the conventional theory has all sorts of adjustable parameters such as evolution, deceleration parameters, space

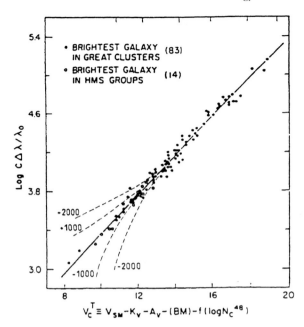

Fig. 9-2. A Hubble diagram (redshift versus apparent magnitude) for clusters of galaxies measured by Allan Sandage. The dashed lines have been added to show the effect that supposed peculiar velocities of 1000 to 2000 km/sec, subsequently measured for some clusters of galaxies, would have on the diagram.

curvature *etc. come up with measured Hubble constants which give an age of the universe younger than the age of the oldest stars!* In contrast, the variable mass theory has no adjustable constants—the Hubble constant depends on only one value, the age of our oldest stars. Nothing can be changed and it gets it right. This is a very important test on which the conventional theory fares very badly.

The Hubble Diagram for Galaxy Clusters

There is even unacknowledged trouble in the centrepiece of the Big Bang theory, the Hubble diagram for supposedly distant clusters of galaxies. The problem comes from the reports of a number of observers that galaxy clusters have peculiar velocities of from 1000 to 2000 km/sec. If this were true, the whole lower third of the Hubble diagram would blow up as indicated in Figure 9-2. Can the classic Hubble diagram measured for clusters, with its small dispersion from the theoretical line, be correct in view of these large, supposed peculiar motions in the universe?

The answer is yes if the redshifts are not due to velocity. If Sandage has measured only very similar clusters which have galaxies created at nearly the same epoch, then he would get very little dispersion from the exact Hubble relation required by the flat space-time solution of equation (1). The investigators who measured clusters of increasingly different characteristics would get higher dispersion in redshifts but these would represent age differences not velocity peculiarities.

This brings us to the thorny problem of Chapter 6. Why do clusters at more or less the same distance, but with different redshifts, such as those associated with CenA as pictured in Figure 6-7, show an even approximately linear Hubble diagram? One could note the fact that newly created galaxies had low luminosities, and that as they aged their luminosity increased, and their redshifts dropped. That would produce a slope in the Hubble sense. If the slope significantly differed from the Hubble slope as it seems to for the higher redshift clusters in Figure 6-14, that would be instant disproof of the redshift-distance hypothesis. But one could argue that the scatter from the Hubble line in Figure 6-14 was so large, even with the scatter reducing properties of a double logarithmic plot, that it is not possible to decide whether it defines, even in the mean, an acceptable Hubble line or not.

But if we take the majority of clusters pictured in Figure 9-2 as belonging inside the Local Supercluster, then we would have to have some mechanism whereby their luminosities increased inversely as the square of their intrinsic redshift. Now we know that the intrinsic redshift varies inversely with the particle mass. The crucial question then becomes: for galaxies born at different epochs, will their luminosities vary as their particle mass squared? The most general answer was given by Fred Hoyle in 1972. In a paper called "The Developing Crisis in Astronomy" he derived the exact Hubble relation for an age-dependent redshift by remarking that the luminosity had the physical dimensions of m^2. (See *The Redshift Controversy* ed. by George Field *et al.*, W.A. Benjamin, Inc., p.299)

As for the empirical approach we note that what is known from the spectra of galaxies in Abell clusters indicates that we are seeing mostly the luminosity of the galaxies as contributed by the stars that compose them. For a galaxy with intrinsic redshift $z = 0.1$ calculations indicate that the oldest stars are only about 12% younger than the 15 billion year age of the oldest stars in our galaxy. Such a difference would not be readily detectable in a composite spectrum. On the other hand the question of whether the stars in a galaxy whose intrinsic redshift was 0.2 would be less luminous by a factor of the ratio of their redshifts, $(1 + .1)^2/(1 +.2)^2 = .84$, or if the same luminosity, less numerous by a factor of .84, is a difficult question to answer at this point. Empirically, if the mass-luminosity relation, L, varies as M^a; then for stars in our own galaxy $2.8 < a < 4$, and for nearby galaxies a very uncertain $.2 < a < 1.4$.

This is the most uncertain point at present in the non expanding picture of the universe where most high redshifts are nearby, young objects. It would therefore be of crucial importance to investigate further the Hubble relation for various kinds of galaxy clusters. Since redshifts for many cluster galaxies are now known, it only requires careful photometric determination of apparent magnitudes to carry out the first step of checking more carefully their redshift-apparent magnitude relations. Clusters like Abell 85 which have sets of discretized redshifts (Figure 8-16) would be particularly revealing.

Cosmic Background Radiation (CBR)

Very weak photons, indicative of low temperature and coming smoothly from all directions around us, were discovered accidentally in 1965. This "CBR" radiation was almost immediately hailed as another, especially decisive proof of the Big Bang. In fact it is very difficult to reconcile with the Big Bang in my opinion. The reason for this is that in an expanding universe radiation from different distances would have different temperatures and the very precise black body curve of temperature 2.74 K which is observed would be strongly smeared out. Because of this it is necessary to restrict the radiation to a very thin shell at the most distant edge of the universe. This shell is supposed to represent the region in which radiation suddenly "decoupled" from matter at some arbitrary point near the beginning (*i.e.* was no longer absorbed and reemitted but flowed freely out into space). For why this shell is so extremely thin, I have never heard a reason.

As measurements continued, the surprising smoothness of the radiation began to worry people. Numerical predictions kept being lowered. The irregularities due to primordial galaxy formation were not apparent. Finally in April 1992 all news media were blanketed with the announcement that a satellite observing the microwave spectrum had detected irregularities in the CBR. There were remarks about Nobel prizes and "having seen the face of God." But it was never explained how something smooth to one part in a hundred thousand could represent a surface where the photons were breaking loose from spaces between clumps of protogalaxies.

Actually this extraordinary smoothness of the CBR seems to be the most important part of the observation. Also it seems to me to be a very strong argument for a non-expanding universe. This comes about because the intergalactic medium can be observed from here to as far as you wish without any velocity smearing due to expansion. The integration through this largest of all possible distances is most capable of smoothing out all fluctuations in background radiation received from all depths of the universe. In the non-expanding universe an obvious, and much simpler, explanation for the CBR is that we are simply seeing the temperature of the underlying extragalactic medium.

What this intergalactic medium might be is an interesting speculation. It used to be stated that the Big Bang *predicted* the temperature of the cosmic background. But a review of the history shows that George Gamow predicted T = 50 K in 1961. It was the static, tired light models by people like Max Born which predicted values around 2.8 K. As early as 1926 Arthur Eddington calculated the photon temperature in and around galaxies as about 3 K. Many investigators have since pointed out that if one takes the ambient galaxy starlight and thermalizes it into lower energy photons (redistributes the energy into an equilibrium state), one gets closely the observed microwave background temperature.

It is natural to think of the "material vacuum" or the "zero point energy field" as possible thermalizing components in intergalactic space. This is simply saying that there is no such thing as empty space—that it contains at least some electromagnetic field and possibly quantum creation and annihilation and/or virtual particles. For example, newly created low mass electrons would be extremely efficient radiation thermalizers.

One specific proposal was made by Fred Hoyle and Chandra Wickramasinghe, namely that iron whiskers were blown out of supernovae. Such whiskers are known to absorb strongly in the microwave region. We would then be seeing the temperature of local space. As Hoyle put it: "A man who falls asleep on the top of a mountain and who wakes in a fog does not think he is looking at the origin of the Universe. He thinks he is in a fog."

But this fog would be transparent at both shorter and longer wavelengths. At longer wavelengths (beyond about 20 cm in the radio spectrum) one should be able to see redshifted fog in an expanding universe. So I prefer the non-expanding universe and a thermalizing agent which is visible over larger distances in extragalactic space. It should be made clear, however, that this is the frontier where new ideas compete and the answer may lie in some unexpected direction.

It is interesting to note that establishment astronomy has poured millions of dollars into just the *analysis* of cosmic background radiation (beyond the enormous costs of the observations). One of the analyzers of this data was describing in a public lecture how the slight irregularities in this astonishingly smooth background was somehow final proof of the Big Bang. (These tiny, irregularly placed ripples are only one hundred thousandth to one millionth of the signal.) A question came from the audience whether the quantization of extragalactic redshifts would effect his analysis. Now recall Figures 8-4 through 8-9, which showed that redshifts were essentially 100% quantized. The answer of the Big Bang theorist came back—"Oh no, that supposed redshift quantization is just meaningless noise riding on top of the signal"!

The Quasi Steady State Cosmology (QSSC)

To illustrate the point that there is healthy disagreement on alternatives even between those who feel that the Big Bang is the opposite of reality, we should briefly discuss the QSSC. In 1993 Hoyle, Burbidge and Narlikar put forward the interpretation that the universe was continually creating itself (steady state) and that the creation episodes were causing it to expand (the quasi part). Actually they had a periodic oscillation in which the universe contracted in creation phases. This was superposed on a longer-term secular expansion. They were able to explain many of the contradictions to the Big Bang model.

One thing they did not explain, however, were the high redshift objects associated with the low redshift objects. (Do not be alarmed: they championed the observational validity of the associations.) Naturally I was wildly enthusiastic about their new matter creation in the presence of strong concentrations of matter. But they created the matter with terrestrial particle masses. If they had only created the matter at zero mass and let it grow with time I felt they would have explained all the redshift anomalies and done away with the need for an unstable expansion.

I also felt unhappy that the establishment would say "Oh they are compromising the steady state and partially accepting a beginning for the evolution." But of course my QSSC friends would feel even more unhappy when the establishment said "Arp is claiming crazy things about galaxy clusters, this proves you can't believe any of the observational evidence against the Big Bang." And then, in the matter of the expanding or non-expanding universe, there would come the inevitable question "Whose side is Narlikar on, anyway?" I could only shrug and say: "He's still researching the issue."

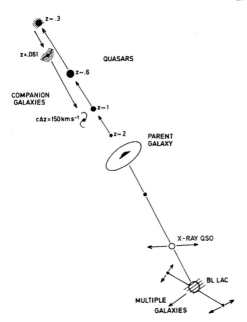

Fig. 9-3. A schematic diagram incorporating the empirical data for low redshift central galaxies and the higher redshift quasars and companions which have been found since 1966 to be associated. It is suggested that the most evolved companion galaxies have relative intrinsic redshifts of only a few hundred km/sec and can have fallen back closer to the parent galaxy.

After all the whole moral of the imbroglio is: The one thing that is certain to be a disaster is to commit early to a shaky assumption and then recruit a lot of people to support it.

The Empirical Model

The greatest mistake in my opinion, and the one we continually make, is to let the theory guide the model. After a ridiculously long time it has finally dawned on me that establishment scientists actually proceed on the belief that theories tell you what is true and what is not true! Of course that is absurd—observations and experiments describe objects that exist—they cannot be "right" or "wrong." Theory is just a language that can be used to discuss and summarize relationships between observations. The model should be completely empirical and tell us what relationships between fundamental properties are required. In an effort to avoid this trap, I want to go back at this point to the observations and summarize the patterns and regularities, which have been observationally established.

Figure 9-3 gives a schematic representation of a large, low redshift galaxy ejecting small, high redshift objects.

How do we know the different redshift objects are at the same distance?
- The high redshift objects are associated with the low with a strong statistical certainty. There are cases of interactions and luminous connections between them. They tend to form pairs across the low redshift object, which unrelated background objects would not do.

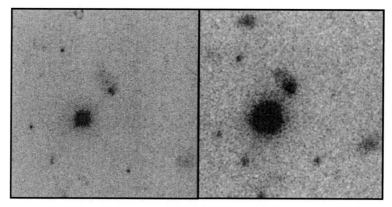

Fig. 9-4. On the left (a) is a Hubble Space Telescope picture of the BL Lac object 1823+56, on the right (b) the same object with the Nordic Optical Telescope. (By Meg Urry and Renato Falomo). It is instructive to note that while the Space Telescope shows better resolution, the greater number of photons gathered by the ground based telescope shows the extremely important, straight luminous connections to the companions paired across the BL Lac.

How do we know they are ejected?

- Since 1948 we have known that galaxies eject radio emitting material in opposite directions from their active nuclei. The radio emitting synchrotron electrons are the lower energy range of the same process that gives the quasars their optical and X-ray luminosity. What mechanism other than ejection could give rise to the pairing of the quasars across central galaxies, which usually show abundant evidence for ejection processes?

How do we know the ejecta evolve into more luminous compact galaxies and finally into normal companions?

- Associations around nearby galaxies show lower luminosity, high redshift quasars closer into the ejecting galaxy (for example M82 shown on page 59 of *Quasars, Redshifts and Controversies*). Associations around more distant galaxies show higher luminosity, medium redshift quasars at greater distances from the galaxy. There is a general change in characteristics from compact objects to companion galaxies which appears to be related to travel time from the originating galaxy.

How do we know that the evolving objects eject second generation quasars which can develop into groups and clusters?

- We see secondary associations around evolving ejected objects, for example, the BL Lac object associated with NGC5548 in Figure 2-3. We see pairs of active objects across BL Lac-type objects (Figures 1-9 and 1-18). And when we look at optical photographs such as shown here in Figure 9-4b, we actually see straight optical connections to companion galaxies on either side of a BL Lac object. (Also see the cluster of quasars around 3C345 in Figure 8-13).

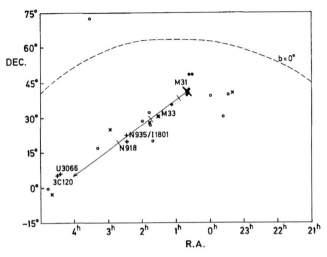

Fig. 9-5. Conventional members of the Local Group (cz_0 <300 km/sec) are plotted as filled symbols. Open symbols (dwarfs) and crosses (spirals) represent all galaxies with 300< cz_0 <700 km/sec. Below M33 additional, higher redshift objects apparently associated with minor axis direction of M31 are labeled. Marks along minor axis direction are at 50, 150 and 400 kpc, just the extent to which aligned companions reached in three independent studies around spirals.

How do we know that the ejected quasars start out with high velocity and then slow down as they evolve?

- The differences between redshifts of pairs of quasars give the ejection velocity for medium redshift quasars as about .1z (Chapters 1 and 2). For the higher redshift quasars, Figure 8-8 shows the spread (of Δz) around the quantized redshift value $z = 1.96$ is appreciably larger. In any case the evolved galaxies show very small peculiar velocities and therefore the ejection velocity is probably lost with time.

Origin of Companion Galaxies as Observed in the Local Group

In Chapter 3 we argued that the empirical evidence of distribution of objects of various redshifts along the minor axes of active galaxies suggested evolution of quasars into companion galaxies. This is schematically summarized in Figure 9-3. But the most conspicuous example of the relationship of such objects could not be closer—right in our Local Group of galaxies which is dominated by M31—and could not have been more deliberately ignored.

Figure 9-5 shows that the major companions are all more exactly aligned along the M31 minor axis than in any other known case. Galaxies of redshifts up to $cz_0 = 700$ km/sec have been added to the normally accepted members of the Local Group. This is necessary because although everyone accepts companions in more distant groups with redshift ranges of over 800 km/sec, they have customarily only accepted members with less than 300 km/sec in our Local Group because that would have made it too obvious that companion galaxies are systematically redshifted. The additional Local Group companions, however, are generally dwarfs and low luminosity spirals and are clearly not background galaxies. They define a most exact line of companions coming out along the minor axis of M31. Apparently this is a case where the direction of the projected minor axis has not moved much in the lifetime of the companions.

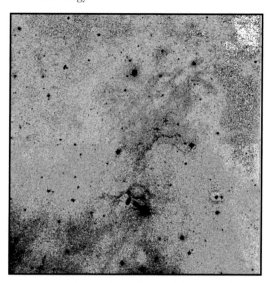

Fig. 9-6. High contrast copy of
103a-E (red) Palomar Sky
Survey print. The galaxy
apparently interacting with the
nebulosity, below center, is
NGC918. At upper left the
interacting double galaxy at
the center of a semi arc of
nebulosity is NGC935/IC1801
(see Fig. 9-5). Field is 2×2 deg.

But the most startling observation is shown in Figure 9-6, where it is seen that along this minor axis alignment of companion galaxies is a string of nebulous clouds which contains higher redshift galaxies. The galaxies noted in the figure caption have redshifts of $cz_0 = 1625$, 4302 and 4434 km/sec. They are, however, obviously interacting with these nearby (Local Group) clouds. The clouds are seen on Palomar Schmidt Sky Survey red and blue plates and IRAS (Infrared Astronomical Survey) maps. NGC918, identified in Figure 9-5, is shown in Figure 9-6 to be ejecting along its own minor axis into the most luminous region of the adjacent clouds. There is also an obviously exploded semicircle of clouds around the disturbed pair, NGC935/IC1801. These are *prima facie* evidences of medium high redshift companions evolving from, or along with, ejected material along the minor axis of M31. *How could all the observatories in the world avoid further observations of this phenomenon after it was published? (Astrophysics and Space Science* 185, 249-263, 1991).

As for even higher redshift objects, Figure 9-5 indicates that the very strong, radio quasar-like object, 3C120, is also along this minor axis line. (See *Quasars, Redshifts and Controversies* p128-131 for more on 3C120.) Since this object is about 700 kpc projected distance from M31 (close to our 690 kpc distance from M31) and is very close to clouds in our Milky Way galaxy, the question arises: Is it a quasar-like object ejected near to our own galaxy from M31? The reference in *Astrophysics and Space Science* above shows also the infrared clouds expelled or illuminated on either side of it, as is also the case with the nearby galaxy UGC3066. The latter galaxy with a redshift of 4594 km/sec is very close to the redshift of the pair NGC935/IC1801 further back along the line to M31. A finishing touch is that the Palomar Sky Survey plates show a long, luminous filament coming down from the north and pointing almost exactly at 3C120 (*Journal Astrophysics and Astronomy* (India) 8, 231,1987). The above reference also shows a nest of high-redshift quasars close around this Seyfert-like object. 3C120 requires an observing project of its own.

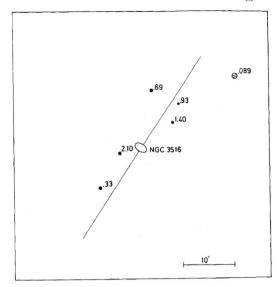

Fig. 9-7. All bright X-ray
objects around the very active
Seyfert galaxy NGC3516.
Redshifts measured by Y. Chu.
Redshifts are written to the
upper right of each quasar and
quasar-like object.

Put aside for a moment that all this is direct, nearby proof that high redshift objects are ejected out along minor axes and evolve into low redshift companions. Just consider these astonishing observations involving our neighboring extragalactic space. Do astronomers really prefer to elaborate obsolete theoretical assumptions rather than make new discoveries?

A Dramatic Confirmation

After this book was finished an electrifying e-mail arrived from Yaoquan Chu, the same Chinese astronomer who confirmed so beautifully the quasars associated with the Virgo Cluster (Figure 5-13) and earlier initiated the confirmation of the quasar redshift quantization. I had seen him at a UN/ESA conference in Sri Lanka and shown him the new X-ray, quasar candidates that had been physically associated with Seyfert galaxies. He was eager to measure redshifts with the relatively modest 2.2 meter Beijing telescope. His e-mail reported the results around the famously active Seyfert, NGC3516.

The numbers could not be better! First of all, as Figure 9-7 shows, the five quasars plus one BL Lac-type object are ordered with the most distant having the lowest redshift, and each successively nearer quasar having a higher redshift. The apparent magnitudes also decreased roughly in this sequence. *This is exactly as the summation of all previous empirical evidence showed in Figure 9-3—a schematic that had been prepared more than a year earlier.* And, of course, it is exactly as required by the variable mass theory when newly created matter is expelled from an active nucleus.

In addition to all this the measured values of the redshift leaped out at a glance as being quantized. The six are listed below against the previously determined values:

Quantization

observed NGC3516:	z = .089	.33	.69	.93	1.40	2.10
Karlsson formula:	z = .061	.30	.60	.96	1.41	1.96

Finally, these six are roughly aligned across NGC3516. I quickly checked the minor axis of the Seyfert and, and behold, *the alignment was centered on the minor axis within a cone of about ±20 degrees!* So with *one object* we had confirmed:

1) alignment of quasars along the minor axis
2) the decay of the redshift and increase of luminosity as the quasars travelled outward, and
3) their evolution into companion galaxies
4) the quantization of the evolving redshift steps.

This news had come when I was in a particularly low point because of the rejection of the paper on further evidence for excess redshifts of companion galaxies— rejected after more than two years at the major American journal and then by the major European journal. Like so many other papers, the referees had made a minor claim which they knew, on some level, to be wrong or irrelevant and wrapped it around with some rude and insulting remarks. The editor had forwarded this with obvious approval and without realistic chance for rebuttal. The part that produced the most anguish was that in most cases I knew these referees and editors. I knew them as acquaintances or even friends. And yet when it came to defending a personal commitment they were ready to drop fairness and principle. That hurt and made things seem unjust and hopeless.

Of course, when the news came through from Chu I talked to Margaret Burbidge, she told Geoff and he proceeded to tell Fred Hoyle. I e-mailed Jayant Narlikar, and when I talked to Geoff on the phone we were all very excited. It just seemed to be an irresistible dawn after a dark night. But at the end of the conversation with Geoff we came to a despondent note: "How can we communicate these important observations?" Always before, I recalled, the senior establishment had encouraged a post doc to "test" the result—and always it had turned out to involve an inappropriate sample. For example, let's not calculate the probability of quasars where they are found but let's calculate the probability for where they *might* have been found! But most effective, these most influential people in the field, in their jovial camaraderie, would simply ridicule anyone who had reported discordant results. How can one fight rumor? I think the only answer is that one must fundamentally change the structure of academic science. Communication must be directly to fellow researchers and the public with no possibility of censorship. This is the major aim of this book. It will take time but all the more reason to start immediately.

Since science is supposed to be characterized by successful prediction—it is significant to note that the most important single observation of quasars being ejected from an active, low redshift galaxy, the just described Chu paper, was rejected without ever being sent to a referee by that leading journal of trustworthy and important results, *Nature* Magazine. This latest news flash reassures us that conventional science is perfectly predictable! It is finally appearing in the 20 June 1998 *Astrophysical Journal*.

The Origin of Companion Galaxies

As shown earlier in Figure 3-27, the alignment of quasars along the minor axes of ejecting galaxies coincided with the alignment and distances of companion galaxies. The results just reported on NGC3516 dramatically confirm this. At the same time this empirical data relieves a worry that had hung over my shoulder since 1968, namely the lines of older galaxies. For example, the E galaxies aligned along the jet of M87 as shown in Figure 5-3. Why had not these older galaxies drifted off this alignment into the general field in all this time?

Now the answer presents itself—being ejected along the minor axis they have no angular momentum and simply remain along their original ejection direction. (HST images show the jet along the minor axis of an inner disk in M87). Only gravitational perturbations gradually increase the spread from the original ±20 degrees, or less, for the youngest to the average ±35 degrees for the older companions.

This result means we now have complete observational information on the evolution of galaxies from the small, high redshift quasar stage through the essentially normal companion galaxy stage. What remains is to explore what information we have on the earlier stages—the stages between creation of an amount of new matter and its formation into a high redshift quasar.

The Earliest Stages of Quasars

One would think that when the quasars were still inside the nuclear regions of a galaxy they would be hidden from view and we would not be able to say much about them. But fortunately radio waves penetrate dust and gas very well so that interferometric techniques (using the resolving power of widely spaced radio antennae) can give very fine resolution of what is going on in the innermost regions. An example of the interior of a radio galaxy is shown in Figure 9-8. *The small condensations coming out of the center of this radio galaxy are only a few thousandths of a second of arc in size!*

Now the Very Long Baseline Interferometry (VLBI) can actually show these condensations to be typically moving outward with speeds of from a few tenths of c to nearly the speed of light. It would seem preposterous to imagine that these lumps are anything else than the proto radio, optical and X-ray quasars which we see moving along ejection lines which eventually reach out to the order of a degree around active galaxies. Immediately this gives direct confirmation that the objects start out fast and slow down as they evolve.

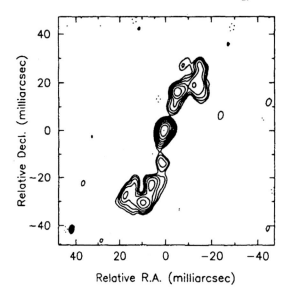

Fig. 9-8. A very high resolution VLBI map at 5 Ghz of a representative radio galaxy (From Wilkinson *et al.* Ap. J. 432, L87, 1993). Note that the smallest radio condensations emerging on either side of the nucleus are, at a maximum, only a few thousandths of an arc sec in size.

I cannot resist an aside here: Toward the end of 1996 when a number of pairs of quasars across Seyferts were appearing in print, *Science* Magazine published a news note with the usual disclaimers by establishment experts. They implied the only observation that could prove quasars nearby was to measure the proper motions across the sky of some of these quasars to see if they were actually moving with the implied speeds at the closer distances. It was clear, however, that pointing accuracies and time baselines would not be adequate for a comfortably large number of years. *But here with the VLBI already existed the tremendously accurate measures on the quasars when they were at the stage of the their fastest travel!* (For example, in Figure 9-8). I recalled sitting next to an old time friend listening to a proposal for an enormously expensive new telescope. "When they run out of money they'll have to think", he muttered.

But there is another strikingly important result to be gleaned from these VLBI observations. The size of the smallest lumps which are being ejected is less than a few milliarcseconds. That means by the time they have arrived at the stage of a medium redshift quasar or BL Lac object, they have grown in size by a thousand fold! That's a billion in volume and the mass density, if anything, has increased. Any useful theory has to explain how something which has come out of such a small nucleus winds up so large.

M87 and Superfluid

In Chapter 5 we saw that there was a giant radio galaxy in the Virgo Cluster variously called M87, Virgo A, NGC4486, as well as 3C274. (It has so many names because it was noteworthy in so many different catalogues.) As early as 1918, a blue spike had been discovered emerging from its center with the 40-inch refracting telescope at Lick Observatory. Plate 8-18 shows how present day radio telescopes reveal luminous knots being ejected along this jet. High resolution with the Hubble Space

Telescope reveals a series of optical knots, some smaller than .02 arc sec (1.4 parsecs at the distance of M87). They are aligned exactly down the axis of this famous jet.

We have known since 1968, however, that giant radio galaxies characteristically have companion galaxies aligned along their radio, X-ray jets (Figures 5-3, 5-4 and 5-5). And we know from Chapter 3, Figure 3-27 and previous sections in that chapter that young quasars are aligned together with these normal companions so that the lines of quasars must evolve into lines of galaxies. They are ejected preferentially along the minor (rotational) axes of the parent galaxy on radial, plunging orbits so they do not wander very far from the original lines. That means that these blue objects coming out of the nucleus of M87 must evolve into quasars and then into companion galaxies. The spectra of the knots are high-energy continua, just like the variety of quasars called BL Lac objects as discussed earlier. The BL Lac objects start to show evidence of development of stars, so we can trace an empirical evolutionary continuity between the small synchrotron knots emerging from M87 and the eventual older galaxies which populate the Virgo Cluster.

Wonderful… But what about the conventional calculations on the jet in M87 involving tremendously complex equations with shock waves, plasma instabilities, twisted magnetic fields, black holes, and so on and on? The mathematics will all have to be repeated with a low particle mass plasma! Why? Because the variable mass theory is the only candidate theory to explain the high intrinsic redshift of the quasars and their rather rapid decay into the only slightly excess redshifts of the companion galaxies. This means, as outlined in the immediately preceding sections, creation of new matter near zero mass and its emerging with near the velocity of light. Initially the plasma particles have low mass and high interaction cross section. A perfect description of a fluid! But this is just what Ambarzumian intuited 40 years ago by simply looking at pictures of galaxies forming by ejection from larger galaxies. He called it a "*superfluid.*"

As time passes, the particle masses in the superfluid grow and the velocities, both systematic and random, must slow to conserve momentum. Therefore the plasma cools and condenses as it evolves into quasars and finally into young galaxies. This would be the Narlikar/Arp prediction. This is precisely where the current theoretical explanation fails on two counts.

1) The observed proper motion of the knots in the M87 jet require the ejection velocity to be exceedingly close to the velocity of light. But laboratory physics requires that particle masses approach infinity as velocity approaches c. Therefore a standard plasma would require an impossible amount of ejection energy. Also the high-mass particles would need to overcome the irresistible pull of a black hole.

2) Knots of standard plasma would have so much heat energy that they would expand and dissipate instead of forming the observed lines of quasars and galaxies.

In contrast, zero mass particles initially come out with the signal velocity, or c, and gain mass. In order to conserve momentum they slow their translational velocity and also their random (temperature) velocities. In other words, the hot plasma cools. *At last—the way to form self gravitating objects!* Since the beginning of the Big Bang and the discovery of ejected lobes of radio plasma the problem of how to form dense bodies from a hot, gaseous medium has been lurking in the closet. Now we can attempt particle formation, element synthesis and hierarchical grouping without the Friedmann/Einstein assumption which requires virgin birth, spontaneous condensation of hot gases and cooling by collision.

Searching for a Better Theory

The most difficult problem for a theory is to explain why matter ejected from an active galaxy nucleus has a much higher redshift than the galaxy from which it originated. The strongest clue, which has been emphasized throughout this book, is that the high redshift objects appear young, *i.e.* in a dynamical and radiation low entropy state, before they have relaxed and run down. Thus, the question becomes unavoidable: "What would newly created matter look like?"

The answer could come logically from the question "How do you operationally define the inertial mass of an electron?" Or it could come from a general solution of an equation expressing the balance of energy-momentum in the universe. Either way the answer would be that you start with a localized potential in space-time and start growing the particle. You don't start out with something having the mass of even a local quasar and pop it suddenly out of a tiny nucleus. Matter seems hard enough to "create." At least you should give yourself the advantage of not having to do it instantaneously.

The outstanding result of this answer is that you cannot avoid high redshift for young matter! Because the younger the electron making the orbital jump, the less massive it will be, and the weaker (more redshifted) will be the emitted photon. Moreover as the particles age, they become more massive; therefore, the ensemble becomes more luminous, rapidly at first, but then more slowly as its light horizon reaches a less dense environment.* As its luminosity grows, its redshift drops, evolving into what we consider "normal" galaxies, *i.e.* like our own. Also as the assemblage ages, its growing mass slows its initially high ejection velocity in order to conserve momentum. The galaxies finish with very slow relative velocities as observed.

This is the kind of theory we are looking for—simple, capable of being visualized—one that can connect together the puzzling observational facts that presently confound understanding. It seems to me that this should be the new working hypothesis that is useful in opening up new directions of investigation until further paradoxes are encountered. We are certainly not at the end of science. Most probably we are just at the beginning!

---

*   If we move all the quasars, young galaxies and X-ray clusters inside the Local Supercluster, as seems to be empirically required, the Local Supercluster will have a much higher density contrast with the rest of the visible universe than now supposed. In that case, we would expect a rapid change in intrinsic redshift for objects up to about 5×10^7 years of age, if that is the diameter of the Local Supercluster, and then slower changes as the light horizon moves outward through relatively empty space.

Mass Creation and Quantum Mechanics

One of the great searches in modern physics has been to connect the realm of the sub microscopic quantum mechanics to the macroscopic world of classical mechanics. There are, however, some classical formulae that seem to apply in the quantum domain if $m^2 < 0$. (See I. Khalatnikov, *Phys. Lett. A*, 169,308,1992.) This treats the imaginary number im as a quantum mechanical variable. It is very provocative, therefore, when the square of the amplitude gives $(im)^2 = -m^2$, a kind of potential mass which can only be realized by crossing the m = 0 boundary.

But in a very fundamental sense, the Machian physics which we depend on to fit the observations—that is what bridges the gap between classical dynamics and quantum mechanics. Because the particle "feels" the mass with which it communicates inside its light horizon, it is in contact through an electromagnetic wave whose particle aspect materializes and dematerializes like a quantum.

Cosmologically, the physics that assumes particle masses constant with time is not valid. What goes on in the rest of the universe affects what happens everywhere else. In addition to the pictures they form in their minds, I think it is very important for humans to realize that the fundamental particles that make up their bodies and brains, and thus they themselves, are in some ill understood way in continual contact with the rest of the universe.

Summary of Big Bang *vs.* Continual Creation

Figure 9-9 schematically summarizes the arguments we have been making that the Big Bang needs to be supplanted by a more rigorous, simpler explanation of the observations. The left hand side of the chart shows that the Friedmann solution started off in 1922 with the dubious assumption that particle masses are constant forever. That immediately led to expanding, generally curved space-time in which all redshifts were due to increased recession velocity with increasing distance. That assumption has led to a head on collision with the observational brick wall which requires extragalactic redshifts to be predominantly not velocity, but instead, age related.

The right hand side of the chart shows that the more general, Machian physics, gives the very simple solution that redshifts are proportional to particle masses and hence to their age since their creation. This immediately leads to a predicted Hubble constant which depends on only one parameter, the inverse of the age of our galaxy, and which agrees with the observations much better than the Big Bang. Then the singularities at mass = 0 and time = 0 which so embarrass the general relativistic cosmology become the necessary mass creation points for the variable mass theory.

Friedmann (1922)	Narlikar (1977)
Special solution • m=constant $$\frac{\mathcal{S}(\tau_o)}{\mathcal{S}(\tau)} = 1 + z$$	**General solution** • $m = m(t)$ $$\frac{m_o}{m} = \frac{t_o^2}{t^2} = 1 + z$$
$$H_o = \left.\frac{\dot{\mathcal{S}}}{\mathcal{S}}\right\|_{\tau=\tau_o}$$	$$H_o = \frac{2}{t_o} = \frac{2}{3\tau_o}$$
• Expanding coordinates • Singularities at $\qquad m = 0$ $\qquad \tau = 0$	• Non expanding Universe (Euclidean) • Creation points at $m = 0$
• $z \equiv$ velocity • distance $\equiv \dfrac{z}{H_o}$ 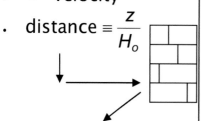	• Quantum $\Leftrightarrow$ classical physics • Merging time scales t, τ • Cascading, episodic creation • Indefinitely large, old Universe
$z = z(t)$ $\longrightarrow$	

Figure 9-9. A schematic summary of the Big Bang (left hand side) versus the more general, variable mass solution (right hand side) of the General Relativistic field equations. The conventional assumption that particle mass, m, is constant leads to an expanding universe and collision with the brick wall of observation that redshifts are not generally velocity but are primarily age related. The Machian solution on the right gives redshift (z) as a function of age (t), predicts the correct Hubble constant, turns conventional singularities into creation points of "new" matter and permits connection with non-local theories such as quantum mechanics.

All well tested local physics is recoverable by making the conformal transformation from cosmic, t time, to local, τ time. We have a possible link to quantum phenomena which is forbidden to the Big Bang, both because the variable mass theory is Machian (non local), and because creation always starts out near m = 0, *i.e.* the quantum domain.

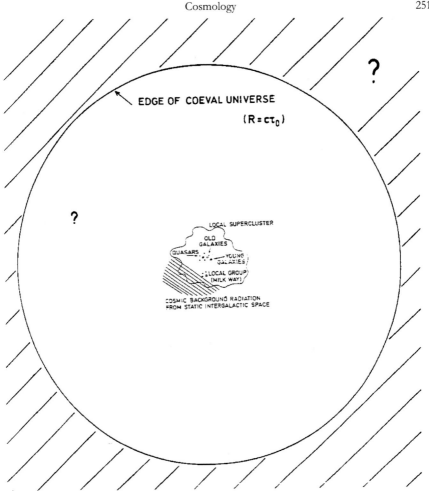

Fig. 9-10. A schematic model suggested by the observations is shown. The region inside an indefinitely large universe within which we can exchange signals is shown as the speed of light times the age of our galaxy. It is like an expanding bubble of awareness in an unknown sea. The intergalactic medium can be smooth and pervasive or slightly concentrated in the direction of the center of the Local Supercluster.

My Best Current Model of the Universe

This will naturally be an empirical model. It connects together what I think are the most important observations in the simplest possible way. Figure 9-10 gives a schematic diagram of some of the main points.

- The universe is not expanding, can be indefinitely large and episodically unfolds itself from many points within itself.

- So far we can only be sure of seeing objects within our Local Superclusters (Virgo and Fornax). The distance to the next superclusters may be very large. We may be seeing only a tiny part of the universe.
- Conservation of mass-energy may apply for the whole, but it is not clear that it applies for the part we communicate with by light photons.
- Patterns in the seeds, which develop into new objects, must be imprinted from very complex laws in this indefinitely large universe. Objects are continually being born and growing but are somewhat different in each generation.

There is one interesting point posed by the boundary in Figure 9-10 which is indicated to be the edge of the contemporary universe. This means for galaxies all born at the same instant, 15 billion years ago (the age of our galaxy), one cannot see any of them beyond this point because that would be before they existed. For younger galaxies, born after this, one would not see them any more after a limit which is closer in.

But for older galaxies, born before our own, it would be more difficult to see them inside the contemporary edge because they have generally diminishing star production. Also beyond the edge, their apparent brightness would be dimmed over their already very faint level by their greater distance squared. So as a practical matter they would have to be intrinsically very luminous if we were ever to see them. An even more difficult point comes, however, if we consider some very luminous, older galaxy beyond the edge. Photons could have started traveling toward us before our galaxy was born and before they knew of the existence of our detectors. Then the question is, if our galaxy did not exist when the photons left the older galaxy, would they register now if they intersected our detector? Regardless of the answer to this question, the edge of our coeval universe is, of course, expanding with the speed of light in all directions. In the Big Bang this is all there is to the universe and it is expanding into nothing. In our model the universe is an indefinitely large substrate into which our awareness is expanding. This means we might experience a surprise at any moment—or eventually.

It would seem that the many new and powerful observing facilities over the world have completely new, challenging studies to carry out if we are to be able to break away from the old paradigm and catch up to the new frontier.

Appendix A

Operational Definition of Mass

$m_p \equiv$ interaction with all particles within light sphere, $r = ct$

$$\therefore m_p \neq \text{constant} \equiv m(t)$$

Field equations from Narlikar (1977)

$$\frac{1}{2}m^2\left(R_{ik} - \frac{1}{2}g_{ik}R\right) = -3T_{ik} + m\left(\Box m g_{ik} - m_{;ik}\right) + 2\left(m_{,i}m_{,k} - \frac{1}{4}m^{,l}m_{,l}g_{ik}\right),$$

$$\Box m + \frac{1}{6}Rm = N$$

spacetime dependent masses—entirely Machian

Reduce to Usual G.R. field equations when m = constant

Note: spacetime singularities in G.R. become $m = 0$ hypersurfaces ⇨ "creation events"

Flat spacetime solution of these equations is given by the Minkowski metric

$$ds^2 = c^2 dt^2 - dr^2 - r^2\left(d\theta^2 + \sin^2 d\phi^2\right)$$

with the mass function

$$m = at^2, \quad a = \text{constant}$$

Mass increases with time from creation of particle.

Mass of all subatomic particles as t^2, the emitted wavelength $\lambda \propto m^{-1} \propto t^{-2}$, hence:

$$1 + z = \frac{t_o^2}{(t_o - \Delta t)^2}, \quad \Delta t = \frac{r}{c} = \text{look back time to galaxy}$$

∴ Redshift varies inversely with the square of distance!
(cosmological redshift consequence of seeing earlier epoch)

Observationally must get familiar Einstein-de Sitter answer because a conformal transformation $\propto t^2$ gives:

$$ds_R = 3t^2 ds \text{ line element in relativistic frame}$$
$$ds_R^2 = c^2 d\tau^2 \text{ where coordinate transformation}$$

is mathematically identical to $ds^2 = c^2 dt^2$

$$t \propto \tau^{1/3} \text{ and } t_o = 3\tau_o$$

t=cosmic time, τ our galaxy time, hence $\tau_o = 15\times10^9$, $t_o = 45\times10^9$ years.

∴ **In a homogeneous universe, for all galaxies which have been created at the same time, we have a dispersionless Hubble relation.**

Appendix B

Operational Definition of Mass

Simple derivation, $m_p \propto \int\limits_0^{ct} \dfrac{4\pi r^2 dr}{r} \propto t^2$

$$\frac{d\tau}{dt} = \text{rate} \propto m \propto t^2 \text{ (relative time rate)}$$

$$\tau = -\int\limits_t^0 \frac{t^2}{t_o^2} dt = \frac{t^3}{3t_o^2} \rightarrow \tau_o = \frac{t_o}{3}$$

To an older external observer, our time τ appears to run more slowly. We appear redshifted. This simple derivation satisfies general case of G.R.! Why? Suggest natural coordinate system is flat, Euclidean spacetime. All physics is in right side of $G_{\mu v} = T_{\mu v}$—can be complex locally—but on a cosmic (homogeneous) scale is simple.

Appendix C

Curved Space-Time?

Curved coordinates are mathematically very complex and help give relativity its reputation for incomprehensibility. But it can be argued that they are essentially mathematical inventions with no relation to empirical physics. For cosmology we have seen they represent complications forced by incorrect assumptions about particle masses.

No curved space-time??? How do you define a point in space? You go x units in one direction, y at right angles and z out of the plane. It is an operational definition. Where does curvature come in? Space is the volume within which such points are defined. To talk of the properties of this space is to attribute properties like that of a gas or solid to an entity defined to be devoid of properties. Or, let's make an operational definition of curved space as what happens to a signal when you send it from point A to point B. The result is conventionally attributed to gravity curving the geometry of space. But what is actually in the space between A and B is electromagnetic waves and particles—there is no substance called "geometry."

As an illustration of how this important point is handled in the media, toward the end of writing this book I read an editorial by one of the more Neanderthal-type political columnists of a leading newspaper. He was recommending to everyone a profound new insight into astronomical advances written by a very good writer and old acquaintance of mine, Timothy Ferris. The columnist admiringly related how this book *The Whole Shebang ...*", *etc.* had rendered in only two or three pages the essential idea of curved space-time! "Ah", I interjected to myself, "But another old friend of mine, the independent-minded physicist Tom Phipps, captured the essence *in one sentence!*"

"Curved space-time I take to be a contradiction in terms."

Primary Reference Frame

Another key aspect of General Relativity is that all reference frames should be equivalent. Work by Franco Selleri [Athens conference 1997 (*Open Questions in Relativistic Physics*, Apeiron, 1998), and *Foundation of Physics Letters*, 10, 73, 1997] and others, however, shows that under the most general transformations of coordinates, the classical Sagnac experiment can only be reconciled if there is a primary reference frame. I feel that this result was logically almost forced by the discovery of the cosmic microwave background. This radiation, supposedly pervading all space, must form a unique reference frame in spite of the fact that arguments have been advanced that it does not contradict general relativity.

Chapter 10

ACADEMIA

The theory that connects together the observations which we have discussed in this book will perhaps always be in continuing debate and development. Of course, given the human spirit of curiosity, it is irresistible to try to relate everything together for a deeper understanding. But it should be kept in mind that we are probably far from any kind of ultimate knowledge. What could be done, and is not done, however, is to use the observations to rule out a 75 year-old model which is presently unquestioned dogma. The mission of academia should be to explore—not perpetuate myth and superstition.

Today any newspaper, science magazine or discussion of scientific funding will take for granted that we know all the basic facts: that we live in an expanding universe, all created in an instant out of nothing, in which cosmic bodies started to condense from a hot medium about 15 billion years ago. The observations are not used to test this model but considerable drama is attempted by implying that each new observation may force an important (but actually marginal) variation in the assumptions of the Big Bang. It is embarrassing, and by now a little boring, to constantly read announcements about ever more distant and luminous high redshift objects, blacker holes and higher and higher percentages of undetectable matter (past 90% it begins to make observations irrelevant). For those who have examined the evidence on redshifts and decided the redshifts are not primarily velocity, however, the important question arises as to how a disproved assumption could have become so dominant.

A Tidal Wave of Elaboration

Some theorists will say, "What's wrong with making a model to see if it works." But in this field the adjustable parameters are endless and one never hears the crucial words: "It just won't work, we have to go back and reconsider our fundamental assumptions."

The practical problem can be appreciated by glancing at any professional journal. One finds an enormous proliferation of articles dealing with minor aspects of models in which the science may be correct but the assumptions are often wrong. Occasionally when evidence appears which cuts the foundation from beneath these heavier and heavier volumes it is almost impossible, in this ocean of print, to be aware of it. But if it does come fleetingly to the notice of an employed astronomer they have a practical choice—to follow up the discordant evidence and compromise their reputation—or continue elaborations of current theory which will enhance their promotion and security. I think the present state of the journals testifies to the fact that the point of no return has been passed.

The Academic Tradition

How is it possible for a scientist to look at a startling piece of evidence—say a nearby quasar of high redshift—and say "Well that is puzzling but I have to get on with my research on distant quasars." I would suggest that this (training rather than learning) starts in grade school and accelerates as the degrees become more advanced. I had only one year of formal schooling up to the seventh grade when I discovered a wrong answer in the back of the book. I was amazed at the reaction of the teacher and the class who could not believe that the answer in the book was not correct. Right from the beginning in science, authority tends to override independent judgment.

When it comes to a degree in advanced research most students earn their fellowships by assisting a senior staff member. Then an advisor suggests or approves a thesis. Finally an exam is given in which correct answers must be supplied. As if these are not sufficient hurdles to original thinking the graduates then face the most excruciating crisis of all, finding a job in the subject they have committed a good part of their lives to, but in a market where there are fewer and fewer opportunities for permanent employment.

Moreover, along this path the most vivid lesson has been that influential professors hold the key to the most desirable positions for those they consider the best students. But what senior faculty consider "best" is usually the research they themselves have undertaken and where they are known for their contributions.

University Departments never go Bankrupt

The ultimate justification for the economic system which the world is currently embracing is that the best way for the consumer to be supplied with what he wishes, is to let the good supplier prosper. If the producer does not make a good or desirable product it is best that he should go broke. But one seldom hears of a department being dissolved with the University saying "Their product is just not good enough, we've had too many complaints or not enough demand." The law of natural selection up to now seems to be suspended for academia.

It is not that scientists are not competitive. I have had reports of lecturers from the most prestigious institutions, in front of large audiences, being asked "What about the evidence for non-velocity redshifts?" With a patronizing smile the answer comes back, "Oh those claims have been completely disproved." One Nobel prize winner confided to an audience of several thousand, "Oh Arp did not get anything right in my course, I should have flunked him but I could not bear to have him repeat the course with me." When I volunteered to give new X-ray results at my alma mater the answer came back, "The committee feels it would not be appropriate for Arp to give a colloquium here." I do not take this personally because they are also destructively competitive among themselves. But I do cry for the science.

Big Science as the Medieval Church

This is far from the first time this parallel has been noted. The church, still in Galileo's time, was the ultimate authority on most important matters. The church hierarchy was handsomely supported by princes and working people, and the life style of the cardinals depended on having people believe that their pronouncements were important and profound. Due to a complex of political, economic and internal events, the church gradually lost power to those who protested.

After the ideals of the enlightenment and the heady rise of astronomy and physics, however, we have the present day situation where all authority on natural law has passed to science. In return for important and profound pronouncements on the nature of the universe the academics are supported with high salaries, expensive facilities, travel, prestige and life time security. They also bestow the power of this institution onto successors of their own choosing.

A Good Press

An unusually entertaining and enlightening example from one of the most respected news services, the *New York Times*, is the following laudatory report of some dialogue between leaders in the field: after one discussant has said that there is no "generic" way in which naked singularities might form according to the known laws of physics, another replies, "Stephen, I am surprised to hear you, of all people, say that. There is one naked singularity that we all agree existed: the Big Bang—the universe itself."

One of the key components of this situation is that academics are generally respected and believed more than other professions in this society. They are trusted to be competent and objective. And while many are—amazingly considering the lack of checks—many others, particularly the most influential, in my experience, are not. I am not maintaining that they are worse than any other segment of society, I am just pointing out that they are perceived to be better. This, of course, is a dangerous situation which tends to fulfill its potential.

From the many comments, communications and manuscripts I receive, it is clear that there are many independent thinkers, in and out of science, employed and unemployed, amateurs, students, retirees. Some are not very knowledgeable, others are very well informed. A range of quality of judgment and ideas from brilliant to crazy are in abundance. But the common theme which binds them together is their increasing annoyance with the arrogance and complacency of establishment science. As one group puts it "a discipline so dead set against reforming from within."

Investigative journalism so far as science is concerned is clearly dead in the water. The media generally take the easy path of handouts and opinions from authoritative sources. No hard work of checking facts and conflicts of interest. In the hopes of stirring some critical reporting from the communications media let me just mention some of the more egregious events connected with science that, in my opinion, have not been deconstructed and therefore have inevitably contributed to its slow downward slide.

The Nuclear Age

At the end of World War II Americans were relieved that the atom bombs had ended the war with fewer casualties in their armed forces than they had feared. Waiting on Treasure Island to go out in the Pacific Fleet, I experienced this relief personally. But there was also a lingering feeling of guilt that so many relatively innocent people had been incinerated without warning. There seemed to be two major currents that developed. One was the hope that the nuclear genie would bring abundant, clean power to the world and somehow atone for its violent entrance. (Of course it would also make a lot of money for the nuclear power industry.) Secondly, the U.S. needed to have an enormous nuclear arsenal so that it would feel more secure than any one else. Both of these goals entailed a lot of experimentation and testing of very dangerous radioactive projects. Scientists were easily recruited to carry out seemingly endless, unwise and harmful schemes; many of which are just now coming to light 40 years later.

Atmospheric nuclear testing was a particularly insane project whereby radioactive elements were rained down on the heads of ill-informed citizens in the name of protecting them. It was not until well-known movie actors like Steve Allen broadcast messages such as: "Mothers, do you realize that radioactive Calcium and Iodine concentrates in the milk which you feed your children?" that public opposition became strong enough to force a halt to the tests. Of course the scientists connected with the *Bulletin of Atomic Scientists*, Committee for a Sane Nuclear Policy and the Federation of American Scientists, *etc.* worked heroically to end testing. The public was more effective, however, because the scientists had a weakness. There were a number of well-known scientists who were supporting the government's claims that the radiation was not harmful.

For example, millions of children were exposed to radioactive Iodine from the Nevada nuclear tests from 1951 to 1962. The average dose to the thyroids of young children downwind of the tests was in the range of 50 to 160 rads compared to 2 rads for people living throughout the U.S. Possible cases of thyroid cancer were eventually estimated at between 25,000 to 50,000 (DOE report cited in IHT 30/7/97). Toward the end of that period, concerned citizens were desperately trying to get the government to release data on radiation dosage from the tests. As a member of a citizen-scientist group in Los Angeles I had headed a group who reported on the biological damage being done as estimated from calculations on the limited data available. A copy of the report landed on the desk of the Atomic Energy Commission. One day I returned from an observing run at Palomar and was told that Glenn Seaborg, chairman of the AEC, had telephoned me and would call back. As a young scientist I waited with some anxiety for this hero-pioneer of nuclear physics to address me. But he never did. The government, and those scientists connected to the government, continued nuclear weapon development and continued to downplay the effects of radiation.

Looking at it now, it seems suddenly clear that specialists such as scientists, the military, politicians and financial people have in common a sense of earned power. They can only conceive of dealing with people of comparable status. They abhor dealing with citizens who have individually little influence. But the fatal flaw, it seems to me, is that people who are interested in power are spurred by emotions which interfere with their reason.

The most frustrating aspect was that the military, and particularly the many involved scientists, knew what harm was being done to people. But everyone also knew that not only would they not stop, but that the information would be suppressed until well after those responsible had retired and were no longer accountable. With poignant certitude this expectation was born out by the fact that a "study" of these events was not commissioned until 1983—*and then it took 14 years* for the report on thyroid radiation doses to be released. From cosmology to pharmaceuticals, it is well justified today that people view institutional claims with skepticism and even hostility. And it is important to always keep in mind who have the vested interests and what they have to gain.

Who can say whether a scientist who has a set of beliefs which coincides with those of politically powerful forces is then rewarded with publicity and money; or whether the opportunity to gain advantages inclines the scientist to see the virtues of the powerful. Be that as it may, there were a number of moderately accomplished scientists, at Universities and elsewhere, who would argue for such things as a "threshold effect." They would say that a well defined relation between radiation dosage and cellular damage suddenly becomes invalid at the point where the current instruments could no longer measure it. In other words, they used science jargon to argue against a probable danger in order to gain a short term goal. But a cell damaging radiation hit is exactly that and arguments do not change the fact. (Excellent summaries of fallout damage and no threshold are given in *Bulletin of Atomic Scientists* Vol. 53, No.6 pp. 46 and 52.)

One scientist with whom I was working on some common projects at the time was effectively educating the public on various radiation dangers. In spite of his Nobel prizes, there was pressure from his fellow faculty members to oust him because he was "aiding the communist cause." The President of the Institute had to issue a warning to the involved faculty to cease and desist. That was the time of the Atomic Energy Commission (AEC) and their traveling exhibits which showed pictures of people basking on the seashore with captions which read "Atomic radiation is no more harmful than sunshine." There was one scientist from an Eastern University who calculated an appalling number of deaths from low level radiation exposure. He even incurred criticism from the *Bulletin of Atomic Scientists*. His arguments seemed reasonable to me, however, and many years later during a discussion of cosmology I asked:

"What do you estimate the total deaths from radiation exposure in that era to be?"

"Nine million" he answered.

The Federation of American Scientists

Because I was a member of, and for some period, Chairman of the Los Angeles Chapter of the Federation of American Scientists in that era, I had the chance to learn how the organization worked. Founded principally by physicists from elite Eastern Universities, they were very effective at reaching, and quietly educating, key government officials about radiation dangers and nuclear policy. They never publicly condemned flamboyant scientists like Edward Teller even during the time of some his most off the wall, wacko schemes such as digging a ditch across Alaska with nuclear bombs.

But the Los Angeles Chapter was a completely different group. With some scientists, but also engineers, social workers, school principals, *etc.* They had the physical principles very well in mind, had good judgment and did very effective work on the West Coast. This included radiation hazards, pollution problems (On the latter we could never get the National group to agree on a course of action.) and other science-community matters. Both groups were admirable in their own way, in my opinion but I was appalled that on the rare occasions when they mixed, that there was instant antagonism. The Los Angeles group wound up getting ejected from the "National" group.

The Los Angeles group did help get started the California air pollution control measures which have turned out to be so necessary. In connection with that I remember visiting the famous "discoverer" of smog in his office at Cal Tech. I was urging his support of a measure to control older cars which were putting an order of magnitude more hydrocarbons out their tail pipes than the rest of the cars. He was strangely reluctant. Finally he took me over to the window and pointed down to the parking lot.

"See that old car over there", he said, "Its mine and I am very fond of it."

The National Academy of Sciences

This is the most prestigious recognition for U.S. scientists. Members are elected by scientists in their own division (but can be blackballed). The Academy is called upon by the government to appoint committees to recommend the best possible solutions to scientific problems facing the nation. One famous and outspoken physicist, however, turned down the honor saying it was a mutually supporting old boy network which did nothing of importance. My observations over the years have noted that their committees often contain members with appalling conflicts of interest in the decisions which are made. Some committee decisions have provoked embarrassing challenges on these grounds.

A personal experience with this system came my way when some years ago the Academy announced a Conference on Cosmology at a large California University. It was instantly apparent that they had invited only the promoters of the current fashion in the field. A number of researchers working with alternative evidence and theories wrote the President of the Academy to protest. I was the only one he answered, apparently because I mentioned money. I said that since there was no critical discussion of evidence at the conference that he could have saved a lot of money by not holding it. In his reply he said he "was particularly proud to have found the money" for this valuable conference in his budget. I was calmly saying to myself that this was the response I expected when I suddenly realized, *"But that was **my** money he found!"*

Electromagnetic Fields

A number of years ago a medical researcher was studying the neighborhoods and houses of leukemia victims to see if there were any differences with the surroundings of non-victims. She noticed that the cases were more prevalent where transformers from overhead power lines came down near the houses. So started the controversy that has lasted for decades about the possible harmful effects on humans of low frequency electromagnetic fields. Of course the inevitable conflicting studies came out, many involving scientists connected with the American Electric Power Institute, but also others, both pro and con. An important division appeared, however, between the empirical observations and the theory.

The National Academy was finally called upon to adjudicate this public health matter. They based their authoritative decision on the theory that human cells had a certain electric resistance and the varying electromagnetic potential across them would therefore produce a current which would produce heat. They calculated that the heat produced would be so minuscule as to pose absolutely no health danger. But then empirical experiments on embryos in chicken eggs showed there *were* effects. Oops! Wrong theory.

A few years ago an unusual newspaper story reported the discovery of very small magnetite particles in human cells. Would the EMF fields slosh these around inside a cell to the detriment of the cell? I have seen nothing more reported on this. But epidemiological studies by Swedish researchers have lent strong support to small but significant adverse effects of EMF fields. And so the counter play between observation and theory goes on.

The aspect that alarmed me the most about this whole situation was the way in which it was treated in the mainstream science journals. Invariably news notes started off with expert opinion as to why there was no credible evidence for danger. (Translation from science speak—we hope this doesn't develop during our tenure). One was left to read between the lines as to what the situation really was. The best information by far came from a series of articles in the *New Yorker* magazine. I was fascinated that this literary, upscale, sophisticated humor magazine would do a more thorough and meaningful presentation than the large circulation science journals.

Aids and Cancer

It is my feeling from the official and unofficial literature that the sudden emergence of AIDS in Central Africa about 1959 was not, in the end, rigorously investigated. Thus was left unsolved the origin of this deadly plague. (See *The White Death* by Julian Cribb, Angus and Robertson, 1996). Are we then prepared to prevent outbreaks of perhaps even more lethal viruses? Do we sufficiently understand the dangers of interspecies virus transmission? To paraphrase a familiar quotation—if we cannot face history we will surely not be able to avoid repeating its mistakes.

Related to this, do we heed the research of a few dedicated scientists who try to communicate the cancer danger of estrogen-like compounds diffused throughout our environment by pesticides, plastics and waste products?

Cold Fusion

In 1989 two chemists, Stanley Pons and Martin Fleischmann, (then at the University of Utah) claimed excess energy was produced when current was passed through an electrolyte containing deuterium and into a palladium electrode. A storm of denunciation broke over them (and others who reported supporting results). The spirit of the criticism can perhaps be best captured by a faculty member of a very competitive, topmost institute. He said, as I remember, "Isn't it interesting that the only schools that report positive results for cold fusion are those with strong football teams." This widely celebrated remark, I felt, contained quite a bit of information when you thought about it.

The obvious point, of course, is that if it were "cold" fusion it would be something new and not behave the same way as "hot" fusion. It would seem to be crucially important to answer the question: Why would scientists dedicated to discovery not simply say, "It doesn't make any difference what you call it let's find out how it operates." Actually quiet research projects are now going in Japan, India, Italy, China and elsewhere. (See *Journal of Scientific Exploration*, vol.10, p.185, 1996). Most recently the Japanese government dropped funding of this research on the grounds researchers could not reliably reproduce their results. Is the "cold fusion" effect a series of mistakes by independent scientists or is it that the essential principal has not yet been discovered? The final decision on hot fusion versus cold fusion may not be known now, but the relative amount of money spent for no results is clearly greater for the former.

The Academics Wander from Fusion to Astronomy

Again in the fusion imbroglio we find the ever-present conflict between the observations and the theory—the cooks and the thinkers. A significant point is that a tremendous amount of money has been poured into research on hot fusion without practical success. For example the "Stellarator" at Princeton was built on the mathematical equations of ionized plasmas with the expectation that a high enough temperatures would be achieved in the closed raceway for energetic particles to achieve fusion. But "instabilities" developed and the beam found many ways to short to the wall of the giant tube before reaching that temperature.

It is not so much that the leader of this project received a Presidential medal for scientific achievements. It is not so much that he went on to successfully promote the Space Telescope project. It is not so much that "the Hubble" was launched with a defective mirror. What is important is that it was too big too soon. What was needed was a wide field optical survey of the dark sky from above the earth's atmosphere (space Schmidt). That would have revealed the crucial relationships of different kinds of celestial objects to each other. We would not now be in the position of looking at exceedingly faint objects in a tiny spot in the sky without the faintest notion what they really are.

The space Schmidt would have cost between 10 and 20 million dollars. The space telescope has cost between 3 to 5 billion dollars. I was one of a group of observational astronomers who spent a lot of time flying to Washington to work out the objectives and design of the space Schmidt. It never had a chance for two reasons: One, it did not cost enough to interest NASA. Two, this same major advocate of the space telescope went out of his way to squash it. I remember a meeting we invited him to in order to hear his objections. He was at the blackboard writing equations that were supposed to show the ground was just as good as space for this project. We interrupted to say that his assumptions about ground conditions were incorrect. He looked around with an injured stare and said, "I didn't know that."

Life on Mars? In NASA?

Decades ago a pair of scientists reported evidence for organic molecules in meteorites. They were decimated by claims of contamination and implications of unscientific conduct. About this time Fred Hoyle was advancing well reasoned arguments and some evidence for widespread life forms in the universe. While stimulating to the public, the discussion caused teeth gnashing among responsible scientists. Without fanfare, however, high altitude flights began to collect micro meteorites with organic molecules. During this period NASA sent three experiments to the surface of Mars to test for life. Two of the three gave positive results, but it was then argued that they should not be interpreted as positive.

In 1976 the Viking probe took pictures of the surface of Mars. As soon as the pictures were released, independent investigators started to analyze them. Since 1979, some investigators have claimed there is evidence that some landforms in one region called the Cydonia Plane may be artificial. One feature bears a resemblance to a humanoid face, and nearby are some possibly pyramidal objects. NASA claims that there is a "scientific consensus" that the land forms are natural. (Different analyses, pictures and conclusions of both sides are admirably summarized in *The McDaniel Report*, North Atlantic Books, Box 12327, Berkeley, CA 94701).

There are some rather sordid events concerning pictures that were supposed to disprove the artifact interpretation but which didn't exist; and then the finding of another that was judged by some to confirm it. There was also the alarming fact of the receipt by NASA of an inquiry into the question of possible social consequences of extraterrestrial discoveries and whether such a discoveries should be kept from the public. Regardless of any individual's estimate of the probability for the objects being either artifacts or piles of rocks, however, it was, beyond measure, the most important object that the next Mars probe could photograph.

Nevertheless, NASA made it clear that even high priority scientific areas were not sure of being photographed again with higher resolution and that the key region of the Cydonia Plane had "no special priority." It was particularly startling to learn that the sole authority to determine not only what images would be released and when, but also what objects would be reimaged, had been given over to a single private contractor. That contractor was an outspoken opponent of the hypothesis of possible artificiality. Some ambiguous reassurances were passed around, and one insider assured me that he didn't *think* that NASA would *avoid* the region.

It was not revealed what the imaging program was for the probe that went into orbit around Mars in 1993. That probe went silent before any pictures were taken. Likewise it had not been stated at this writing what the program is for the Mars probe which was launched in November 1996 and arrived in the summer of 1997. On 26 March 1998, NASA announced: "Mars Global Surveyor to attempt imaging of features of public interest."*

In view of all this, it was rather bemusing to see some NASA scientists call a press conference in the summer of 1996 to announce probable/possible very small bacteria on a rock which had been impact blasted off Mars about 4 billion years ago and landed in Antarctica some 100,000 years ago. No mention of similar small, possible bacteria found by the German scientist Hans Pflug in the 1970's in a carbon-rich meteorite thought to be from the Asteroid Belt.

One lesson from all of this, which seems obvious, is that scientists have to be absolutely honest and straightforward with the public, the people who are paying their salary. Their primary moral obligation is to report the facts and make available a range of interpretations. They have no paternalistic excuse to guard the public from "misunderstandings" or "alarm." If they cannot explain a matter so that a non-specialist can understand it, they don't understand it themselves and they should not cover up this important situation.

Plate Tectonics

As is well known, Alfred Wegner in the 1920's pointed out that geologic features of the West African Coast would accurately line up with similar features in the East Coast of South America when the two continents were fit together. This quintessential piece of pattern recognition drew ridicule and derision from established geologists who, if pressed for a comprehensible reason, would argue that the continents could not drift about because they were anchored in basaltic rock. It is rather startling to see a short

---

*   On 7 April, 1998 NASA released to the press a number of the original face pictures and a view of what appeared to be a low hill with remarks that they had said all along the feature was natural.

time after Wegner's death that the fashion changed so completely as to have all the continents splitting away from a single land mass and go floating debonairly out to sea.

The most compelling piece of evidence, of course is the mid-Atlantic ridge running from the far north to the far south between the Euro-African and American continents. That seam has been *measured* to be opening up a few centimeters per year and spilling material from the interior onto the surface. This is just about the right rate, during the order of magnitude of the age of the earth, to account for the distance that appeared between the continental plates since their breakup about a quarter billion years ago. This has been interpreted by a number of people as meaning the earth is expanding in size and pulling the continents apart. (A notable proponent has been the geologist from the University of Tasmania, S. Warren Carey. But it is best to consult the papers contributed to the Olympia Conference on *Frontiers of Fundamental Physics*, Plenum Press 1994).

One crucial point comes in the Pacific where current conventional wisdom says the American plate is over riding (subducting) the Pacific plate. Carey cites evidence that it is not. In any case, the rim of fire in the Pacific seems to be a region where hot material is rising from deeper regions. Without arguing the details, if the earth were expanding, the continents would have to be moving away from each other. Ironically, they could stay anchored in basaltic rock as was originally so strongly believed! What else besides surface expansion could be a natural explanation for their movement? As for a reason for the earth to expand, if new matter is created in the presence of dense older matter, a small rate of production at its core during the 4.5 billion years of the age of the earth might be a candidate.

Of course, further measurement and analysis is the only scientific way to settle the question—particularly in the region of the Pacific rim of fire. The aspect that astounds me, however, is that the observational and logical difficulties of the current theory are never mentioned, nor is the alternative theory mentioned in texts, media or academic communications. I first heard of it many years ago in a self published book by someone known to me only by the name Sam Elton. But next time you travel on a transcontinental flight you can stare at the map of the Atlantic ridge and Pacific trenches and mull the problem over—expanding surface or random drifting about? One could also ponder its amazing suppression from discussion.

The Restless Earth

In July 1997 a rather startling announcement was released. Researchers claimed that magnetic fields in rocks showed that there had been an abrupt change in the poles of the earth's rotation—that about 550 million years ago the rotation poles had switched to points on the previous equator in a time span of only 15 million years. They claimed "violent earthquakes had been tearing continents asunder, slamming them back together, throwing up towering mountain ranges"—even moving North America to its present position from near Antarctica.

My first reaction to this was that of a typical, conservative academic: "What an irresponsible attempt to garner publicity. They know perfectly well there is no way of moving sufficiently large masses around inside the earth to cause such havoc as changing rotation to a completely different direction!" Then a second thought gave me pause: "They know there is no plausible explanation, yet they had the courage to report their observations anyway." It started me wondering what the power balances in that field were, and what would be the subsequent developments.

But I was then distracted by the even more interesting realization that there was a possible explanation for such upheaval in the interior. Mass creation! But the observations require a *sudden* change. At that point I remembered the discussion in the Olympia Conference (see previous section). The expanding earth advocates were worried because certain measurements indicated that the expansion was going too fast. I reminded them at that time that the mass creation was *episodic.* All the lessons in cosmic evolution we have reviewed in this book have pointed to quantization on all scales with rapid evolutionary jumps between them.

The recent magnetic record evidence is science as it should be—observational evidence presented in spite of the fact there is no currently believable cause. (Although magnetic reversals in the archeological records have been known for a long time without much discussion ventured.) In the recent reports, however, another empirical fact has been connected, namely that in this same Cambrian period new types of animals appeared at rates more than 20 times normal. The explanation proposed was that evolutionary innovations are more likely to survive in small, broken up, isolated populations which resulted from the upheavals in rotation. But regardless of whether this explanation is the correct one, the important point is the coincidence between the two extremely unusual events. This strengthens the empirical validity of both. Again this is empirical science, connecting events together which give supporting but different information on an important but unknown process. This process may be far beyond our imagination at the moment, but this is the only way we could ever approach eventual comprehension of it.

Another point which bears on the internal upheavals in the earth is the exploding planets hypothesis. The evidence is presented well in Tom Van Flandern's book *Dark Matter, Missing Planets and New Comets,*" North Atlantic Press. Since the asteroid belt almost certainly represents the remains of a broken up planet, the empirical evidence for such a process is very strong. The *Meta-Research Bulletin* edited and produced by Van Flandern is also a valuable source of science news of the kind routinely repressed elsewhere, as well as exposition of the editor's ideas about the nature of gravity and other topics in astronomy and physics. The web page www.metaresearch.org is also now open in connection with these matters.

The Gaia Hypothesis

Somewhat related to the question of geological evolution of the earth is the proposal by James Lovelock that the earth might be considered as a living (organically

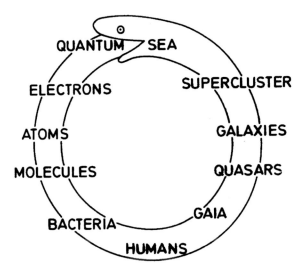

Fig. 10-1. The Uroboros, ancient symbol of the universe as a snake with its tail in its mouth. Major entities are ordered in increasing size. Every part of this hierarchical structure communicates with every other part by means of electromagnetic waves. Its symbiotic nature and evolution now comprise the most challenging frontier.

evolving) entity. The interesting aspect of this concept is the analogy between the huge numbers of bacteria and viruses that inhabit the human body and the humans that inhabit the earth. The bacteria, though a very successful life form, would probably have difficulty grasping the operational purposes of the humans they inhabit. By analogy, humans might have great difficulty recognizing intelligence in much larger organized entities than themselves.*

The pertinence of the Gaia hypothesis to the astronomical observations discussed in this book is that for the first time we have hard observational evidence for the evolution of different forms of organized extragalactic objects, the birth and maturing of younger objects into older objects. Perhaps most important of all we have the beginning of evidence of how matter materializes from the "diffuse" state of the cosmos. We do not know that it returns to an all-pervading state—but it may through the decay of elementary particles. In any case the various bodies in the Uroboros (ancient symbol of the universe as a snake with its tail in its mouth) shown in Figure 10-1 are arranged in a continuum of size. (Some important concepts were quite logically induced in distant times.) Now it may be fascinating to consider how much evolutionary or symbiotic connection there is between various components, but it is clear that all are in continuing communication with each other by means of electromagnetic waves such as photons, machions and various quantum mechanical aspects of the universe.

Creationism

One of the crusades of academic science is against religious creationism. Periodically there arises a messianic need to save the general public from the ignorant

---

* It is interesting to note that Lovelocke's fundamental discoveries of trace constituents of the air and his development of instruments to measure them, which enabled the whole ecological movement to go forward, was not enthusiastically supported by the U.K. science establishment. Eventually he relinquished his tenured position and moved to the country where his children could grow up "seeing the Milky Way." His comment on his experiences was essentially, "Well you have to realize they are not scientists."

beliefs that humans were created in their present form some short time ago, say 8,000 years or so. Activists try to convey the facts of evolution over millions of years as testified by the fossil records and Darwin's theory. They pit the scientific evidence of evolution against the primitive superstition of creationism!

They should blush with shame. Their establishment science is the most blatant possible form of creationism. The claim is that not just humans, but the *whole universe* was created instantaneously out of nothing. So there is small debate about time scales, but the principle is carried much, much further in the Big Bang. The religious creationists are not slow; I have read in one scientific journal that scientists should not try to debate them because they are clever at confusing the audience!

Referees

Refereeing, or "peer review" as it is rather pompously called, is now unworkable. It has increasingly shown that it lets in the bad papers and excludes the good ones, exactly the opposite of what it is supposed to do. Just in abstract principle, science is supposed to be a competition of ideas and indeed, as we have seen, it is very competitive. Is it reasonable then to send your ideas and data to an anonymous competitor who can with impunity often steal, suppress or ridicule them? What happens to the hallowed principle of jurisprudence that one has the right to confront one's accuser?

As an example of a more temperate but nonetheless cutting analysis, we have a paper by David Goodstein of Cal Tech which was actually printed in an establishment journal. The following excerpt is from *Science* 825, 1503, 1992:

> *"The referees must therefore make an ambiguous, not entirely scientific, judgment in a high-stakes game in which the authors are usually known personally to them and are often competitors. Furthermore, the referee knows the editor will not understand the technical details of the report that will be written. If the judgment is wrong or unfair, only the author will know, and the author will not know who wrote the report. The referee can count on the editor's protection and support even if the review is guided by self-interest, professional jealousy, or other unethical motives, because the referee's unpaid help is essential to the editor and the author of a rejected manuscript has an obvious motive to be disgruntled. Referees are never held accountable for what they write and editors are never held accountable for the referees they choose. For all of this to work, the referees would have to have impossibly high standards of ethical behavior, but nearly all referees have had their standards corroded by themselves being victims of unfair referees' reports in the past when they were authors. Any misconduct that occurs under these circumstances is certainly committed by the referee, not the editor, whose behind is well covered. Nevertheless, the editors have managed to create a system in which misconduct is almost inevitable."*

Lest it be objected that most referees are principled and fair you should look through the folder of referee reports that most scientists collect during their career. Some are. In some fields almost none are. I am not just judging this on my own voluminous folder but on those reports to people who are unquestionably competent scientists. Many reports read like an emotional session of psychotherapy—manipulative, sly, insulting, arrogant and above all *angry*. A sample of these should be published because it would allow people to evaluate the objectivity of the information they are being allowed to read. Their best use would be to enliven the ends of controversial articles with short replies from the authors.

In the beginning there was an unspoken covenant that observations were so important that they should be published and archived with only a minimum of interpretation at the end of a paper. Gradually this practice eroded as authors began making and reporting only observations which agreed with their starting premises. The next step was that these same authors, as referees, tried to force the conclusions to support their own and then finally, rejected the papers when they did not. As a result more and more important observational results are simply not being published in the journals in which one would habitually look for such results. The referees themselves, with the aid of compliant editors, have turned what was originally a helpful system into a chaotic and mostly unprincipled form of censorship.

I would propose that there were two obvious principles of scientific communication:

1) Publish all sides of an issue.
2) When there are differences of opinion, the author has the final decision on what he wishes to say.

Editors routinely violate these primary principles. The great rationalization of course is that "You can't let crackpots into a respectable journal" (one minority resident will ruin the whole neighborhood). The situation seems more or less irreparable today. Perhaps it is inevitable that the enterprise has become too large and must fraction into alternative journals—let the fittest survive in spite of establishment subsidies!

Culture Wars

Cooperating groups are efficient at carrying out their own programs but they are hard to redirect to goals which are beneficial to larger groups of which they are a part. One of the most conspicuous examples of this is the hotly debated subject of race and I.Q. A few years ago this pot was stirred anew by the academics Herrenstein and Murray. The obvious implication of their book was that intelligence was inherited. The perhaps not so obvious fallacy was that I.Q. measured intelligence.

The earnest constructors of these tests are not least among its supporters as a culturally unbiased measure of pure mental ability. However, I fall back again on personal observation. I remember about the age of 13 taking this delightful test in which I was asked all sorts of questions which I knew the answers to because: 1) In all the

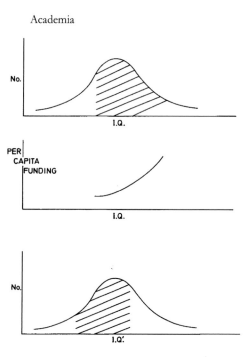

Fig. 10-2a. A schematic distribution of intelligence quotients (IQ's) in, for example, universities or academic science (shaded portion). 2b) A sketch expressing the fact that per capita funding increases with the higher IQ's which elite institutions can attract. 2c) A different test which might measure Insight or Innovation Quotient (I.Q. prime) might be distributed differently in these same institutions but be supported by the same per capita funding relation.

previous time I had not been going to school, I had read widely in adult books. 2) The subjects included subjects discussed passionately in my artist family. I remember thinking, "Boy, the people who made these tests are interesting, not at all like the people in my small town!

We can refer to Figure 10-2a which will drive the academic scientists wild. It shows the famous bell curve of I.Q. distribution but the axes are not quantitatively labeled. This is even more true of the curve below it. Required procedure is to spend a lot of time measuring the exact percentiles or making complicated per capita calculations from exhausting financial statistics. By then the point of the whole result is lost in the details. But in reality most people know very well from general observation that if we are discussing universities or academic science, for example, that the highest I.Q.'s in the society will be more highly represented. Moreover we know that the richest universities will be able to hire the brightest minds and they will be the relatively most richly paid and supported with the most expensive facilities.

Now suppose we devise a new test which we call I.Q. prime. This test measures the ability of people to make fundamentally new insights, to make innovative new solutions to problems. (The ultimate operational definition of intelligence must be that which most promotes the survival of the species. Since we do not know what that is from any given point in time, we can not devise a perfect intelligence test. But from this perspective we can certainly question some of the currently accepted criteria—for example, like the ability to build a nuclear "device"). If we are to believe any of the points discussed in the present book we can construct the bottom curve in Figure 10-2c in which the academic scientists would have some diminution in numbers toward a higher Innovation Quotient. This is simply because of their selection on the basis of current cultural values they are less able to break the paradigm to achieve fundamentally better solutions. *But perhaps the most important conclusion is that the innovative or insightful I.Q.'s are relatively poorly funded or heeded.*

Creativity in Academia

New insights into the relationship between human beings and life processes have come from individual artists and movements. It would be ludicrous to imagine a da Vinci, van Gogh, Corot or Duchamps producing their work as a member of a University Art Department. Writers who move the culture say that the surest way to kill writing ability is to work in a Literature Department. Even seminal workers closer to science such as Galileo, Freud or Gropius would obviously stick in the throat of an academic institution. So why can great physics and cosmology only be produced at a lavishly funded institution? The answer is that it isn't—which is the whole point of the preceding book.

Is the solution to distribute some of this funding on non academic scientists in the hope of encouraging some unheralded genius? Hardly, because how would the award committee be chosen? In the science fields which I happened to notice, the MacArthur Fellowships were mostly awarded to Institutional favorites. The answer is again that we do not know what is right, but we do have pretty good indications of what is wrong.

Societal inertia being as strong as it is, we, or any subgroup, probably could not move it very far in any particular direction even if we had a plan as to where it should go. But we can recognize where it is carrying itself—inexorably toward *defunding* of elite theory organizations. The concentration on ever more expensive hardware cannot save science from a senescent theoretical foundation. The question of how to rescue some of this funding and distribute it to new and innovative research is the most difficult question. It obviously must be more democratic in spite of the individualistic instincts of intellectuals.

Science and Democracy

One of the most self-evident principles which I heard voiced along the way is that in science "You can't vote on the truth." No matter how many people believe something, if the observations prove it is wrong, it is wrong. But as is often the case with humans it turns out that a lot of scientists actually believe exactly the opposite. So many fine, gentle colleagues of mine have said, "Well that evidence looks pretty strong, if you could only get some more people on your side, prominent astronomers, some opinion leaders, to endorse it." *It needs to be accepted.* As soon as they say that, they wonder if something is wrong with the evidence.

But then the other side of the two simultaneously held, contradictory beliefs comes into play. When it turns out that a large number of renegade specialists and amateurs believe contrary to the most prestigious experts, the latter say, well science is not democratic, it is *what the people who know the most say*—that is what counts!

I finally stumbled onto what was going on here during the last presidential election. Everyone was complaining that the candidates changed their stands on every subject with every opinion poll.

"No integrity", was the cry, "what the country needs is some candidates with leadership."

"Wait a minute", I thought, "Isn't this what we have been trying to get for such a long time? Finally a real democracy where the elected do exactly what the electorate wants." But of course, the public at large is not knowledgeable, it would be dangerous to be governed by the unenlightened. This is the self-evident argument that usually closes the discussion. But if you think about the really bad trouble that groups have gotten into, it is almost always because a strong leader has led them into disasters. So we are ultimately forced back to the old homily: "Democracy is a bad form of government—but it is better than any alternative"! As far as science goes it is necessary to be suspicious of everyone, but particularly of the experts. (The operative definition of an expert being someone who doesn't make small mistakes). *Everyone must make up their own mind on the basis of the evidence and the experts should not be allowed to control the presentation.*

Essentially, I believe that competition inside a peer group of specialists will produce a non-democratic structure. As in art or literature the communication should be between individuals and the society as a whole. The test of those communications which truly enlighten and inspire other individuals in the society will be whether they are supported. For radically ground breaking ideas, as always, it will be necessary for the originator to gain a double perspective, which includes communicating and supporting himself as part of the society. Hopefully the present oligarchy of incompletely separated academic church and state would continue to develop toward a democracy of individuals.

Public or Private?

Many universities are primarily supported by private endowments. But the public contributes important amounts through state universities and government contracts which are administered by universities. In this sense academics are like a rare remaining form of the old time guilds. Of course many universities have large investment portfolios and are like businesses run from a powerful corporation office. Their success greatly depends on their public relations with a complex mix of students, alumni, trustees, government and community from local to international. They bear *veritas* on their escutcheon but it is a question how much time they have to nurture it in their heart.

In scientific research the examples we have discussed in this book seem to show how this complex institution encourages the least useful aspects of scholarly isolation while at the same time encouraging the most damaging aspects of competitive pressure to conform to fashionable (largely self promoted) paradigms. The question arises then, are there better ways of organizing research? Private, specialized "think tanks" spring to mind. Perhaps that is an answer if they are sufficiently funded by goal directed money.

An example of such was a department of an institution I was once part of. It was originally founded to build telescopes and explore the universe. Edwin Hubble, George Ellery Hale, Walter Baade and many other astronomical pioneers used the best telescopes of the day to report new findings about galaxies and astrophysics. Their era inevitably passed, however, and the newer staff members competed in emulating accepted concepts. When I was faced with a directive to renounce observations of new phenomena, I chose early retirement. It is interesting that when one of the staff from that period retired as director he stated:

> "The real life of the Observatories results from the free choice of the individuals who use the facilities... Our tradition of free choice has continued to be cherished..."

So, as in many human activities, people often think they are doing one thing while they are really doing the opposite. The problem with this once leading research institute is that it tried to be just like, or perhaps even more like, all the other topmost university departments. The lesson I draw is that truly creative, ground breaking, private research institutes should be kept small and chartered to do primarily what other institutes do not do.

What's Next?

Whenever I go to science conferences these days I hear all around me, "Where is the funding coming from?" "The budget has been cut." "There is no money to hire young scientists." "Positions are being cut back." Everyone is worried. Of course, instead of repressing research and debate on alternate cosmologies the dons of academia could permit meaningful controversy. Allowing people into the excitement of the most fundamental questions of their existence would certainly enhance support of their projects.

After complaints about funding come descriptions of new satellite hardware, ambitious and expensive experiments, telescopes, detectors in progress. There is great pressure to build ever more technologically advanced projects. This pressure comes also from a commercial-engineering society that wants to work, develop new industries and make money. But the supposed goal of all this is to produce new knowledge. If the data is hijacked at the last moment by a group with a need to control beliefs, the whole enterprise is a failure. So the most important people of all to fund are the independent (at present, by default, non-academic) researchers who can communicate all the data and in a form where it can be understood and debated. That this is not presently possible is the insoluble problem which I personally think will cause the whole juggernaut to inexorably decline and regress for a long time before getting back on a useful track. Sadly, I do not see a sufficient residual of innovative academic scientists to reform the institution.

But in addition to this other voices now ask: "With so many poor people in the world, many ill, some starving should we be spending so much on curiosity in the first place?" My feeling is that, while seemingly abstract, effective investigations of the fundamental nature of matter are probably the most practical commitment humanity could make. If humans survive for a long time they will inevitably encounter potentially lethal events: rogue asteroids capable of wiping out the earth, evolution of the Gaia environment, supernovae, passing through vastly different galactic environments, unpredictable events that the animal world regularly encounters and notes by the extinction of another species.

Standing in a modern airport one could easily imagine departing for another planet in our solar system. But to another star, or part of the galaxy—we are prisoners of our finite lifetimes and the speed of light. Nevertheless if humans can guard against moderate catastrophes long enough to have a really long future, who can foretell the possibilities? For example, if it turns out that mass is primarily a phenomenon of frequency, that means we might affect it by subtle wave and resonance interventions. If we live in a Machian universe, the atoms in our bodies are in communication with the far universe. If our matter was materialized from a previously diffuse state we carry the information of an enormously complex pattern that is somehow connected with everything else. In the long future I cannot help believing that knowledge will not only determine whether we survive; but if we do, more importantly, in what direction we evolve.

The Zen of Research

As I was leaving the cosmology conference in Bangalore in 1997 a young Indian couple approached me and asked whether they could speak to me that evening. After they arrived we talked for more than an hour and they told me they were finishing advanced degrees in physics and astronomy and they wanted to do research on the kinds of phenomena that I had reported.

I felt a responsibility to tell them of the difficulties—of the cases where the most talented and hard working young astronomers I knew were forced to leave the field because they were felt to be too open minded about fundamental assumptions—how even conforming to work on fashionable subjects was not likely to ensure a job in a field with declining financial support.

As I went to bed that night I felt terrible. How could I have been so discouraging? They only had the simple wish to investigate some new and interesting phenomena. Was it really impossible in this society? Well, I tried to tell myself, if it was true love you committed yourself anyway and dealt with the problems as they arose as best you could. So I hoped they cared enough to try; but, sadly, I had to do my best to inform them of the reality ahead.

That night I dreamt about a story I had read long ago about a young man who wanted to become the best swordsman in the world. He sought out the world's most renowned Zen swordmaster and asked whether he could be his pupil.

"Allright", the master said, "but you will have to move into my house and do everything I say."

So the young man moved in and was assigned the most arduous tasks in the house—gathering wood, cooking, washing, cleaning. After more than a year of drudgery without a single word about swordsmanship, one day the acolyte was down on his knees scrubbing the floor. Suddenly the master sprang out from behind a pillar and hit him a ferocious blow over the head with a broom handle.

"Master," cried the student looking up from the floor in pain, "why did you do that?"

"That was your first lesson in fencing", the master replied, "—always be on your guard."

When I first read this story my progressive western reaction was that this poor fool of a student was not only allowing himself to be used and exploited but, in addition, to be injured and humiliated. But without taking an iota from his need to face reality and exercise his own individual initiative, I also realized that there was a deeper implication. Anyone who wished to achieve his goal had to be committed enough to deal with and learn from injustices and setbacks. I felt a little better about the previous night's conversation and felt the hope that if enough people tried, eventually the shadows on the bottom of the pool might turn into a better reality.

Final Thoughts

It is clear that no matter how ill we think of the present, ponderous cultural establishment, we are not going to move it precipitously. Perhaps that is fortunate. Deep thinking friends argue that the greatest danger humanity faces today is too rapid technological advance. Before we have a chance to learn of the possible disastrous effect of one change we are already airborne in the next leap. The same undoubtedly applies to societal changes.

That does not mean, however, that we have no responsibility to try for fundamental, lasting change—or that we should go on supporting useless or harmful structures. Inertia will do that well enough. We can concentrate on fighting for support of new and better understandings that will lead to a better way of accomplishing our goals. Private or public? Individual or group? It's a great adventure.

And in the end I must admit, that though my criticisms are passionate, I am at heart an academic who dreams of what might have been and (still the fatal innocence?) might yet be.

Epilogue

As this book was being prepared for printing some new evidence turned up. It was so compelling, and offered such revealing lessons as to why previous evidence had been ignored, that it clearly demanded inclusion as a final summary. It is particularly fitting that just a few pictures enable the underlying observational facts to be grasped at a glance. Then the imagination can leap forward with renewed confidence to the many interesting implications which have been discussed in the preceding book.

The Scintillating Quasar

Over the telephone Geoff Burbidge informed me that observers in Australia had measured astonishingly large variations in radio wavelengths in less than an hour in a quasar. This marked the size of the energy emitting region as less than a light hour in diameter. At its redshift distance the luminosity of the quasar was so enormous that it made the surface brightness incomprehensibly large.

Geoff said: "Chip, that quasar must be closer, find out where it comes from."

So I looked. The first thing I found out was that there was another quasar which formed a strong, flat radio source pair with the scintillating quasar. Then I discovered that the brightest Seyfert galaxy in this whole region fell midway between this pair. (Figure E-1.) The next piece of information was a real shocker! *This Seyfert was emitting enormous amounts of X-rays.* Despite its relatively modest apparent magnitude of V = 15.4 it was one of the 5 or 6 brightest X-ray Seyferts in the sky (at a prodigious 4000 counts per kilosecond).

The other quasar in the pair was also emitting unusually strong amounts of X-rays, 226 cts/ks. The scintillating quasar, although more modest in X-ray emission (40 cts/ks), was a rare emitter of even higher energy gamma rays.

The upshot of all this was that here was another pair of quasars ejected from an active Seyfert galaxy. But the extraordinary nature of all three components ensured there was negligible possibility of the association being accidental. In fact the apparent brightness of the components and the somewhat larger angle subtended on the sky enabled the configuration to be compared to the previous Seyfert associations discussed in the early chapters of this book and a relative distance from us estimated. The distance was less than half that of the previous cases which were primarily at the distance of the Local Supercluster. But this was less than one thousandth the conventional redshift distance. That meant the luminosity had to be less than one millionth of the

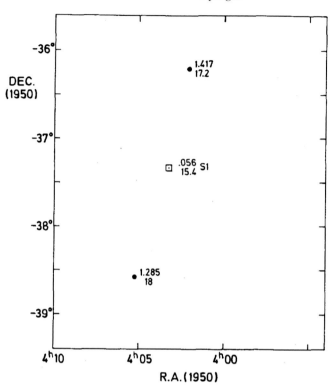

Fig. E-1. The quasar at z = 1.285 is the "scintillating quasar", PKS0405-385. The quasar at z = 1.417 is PKS0402-362. The brightest Seyfert galaxy in the field falls between these two quasars and is emitting the enormous X-ray flux of 4 counts per second.

conventionally assumed luminosity and reduced the embarassingly large surface brightness by the same factor.

Even with this reduced surface brightness, normal physics tells us we have to be looking at a jet directed almost exactly at us and boosted to a velocity extremely close to the velocity of light. As remarked previously, boosting matter so close to the speed of light requires enormous energy. Even reducing these energy requirements by a factor of a million with the closer distance it is difficult to account for the high surface brightness of the scintillating quasar. It suggests that near zero-mass matter flowing out at near light speed (signal velocity) might be required to explain this startling observation.

Two Radio Survey Fields

In 1984 the Westerbork radio telescope surveyed 9 fields, two of which are shown here in Figure E-2. Optical photometry and some spectroscopy was done in 1985 and a quasar of z = 2.390 was discovered near the edge of the Hercules II field. Recently observation with the Hubble Space Telescope has revealed a large number of quasars (5) and galaxies (14) all between z = 2.389 and 2.397. All of these are in a very small area. (The 5 quasars are shown as small dots in Figure E-2).

Is there a large active galaxy nearby which would give rise to these high redshift objects, as we found in earlier chapters here, and in the earlier book *Quasars, Redshifts and Controversies?* Well yes, there is a V = 16.5 mag. galaxy with great plumes coming off a

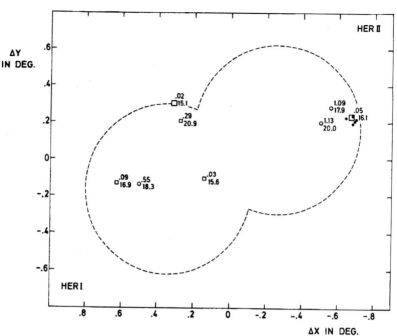

Fig. E-2. These two Hercules fields were searched for radio objects with the Westerbork telescope. Blue radio galaxies are indicated by small boxes. Open circles represent radio quasars. Redshift and apparent magnitude are written above-right of each symbol. Small filled circles represent 5 quasars of redshift z = 2.389 to 2.397.

body broken into at least three distinct pieces. As the square symbol in Figure E-2 shows it is the only blue radio galaxy in the Her II field! Figure E-3 shows a picture of the object taken by William Keel. How could the investigators have missed its significance—only about an arc minute away from this extraordinary cluster of high redshift objects?

As we have learned from the previous evidence, lower redshift quasars are usually found further away from the ejecting galaxy than the high redshift quasars. So looking for catalogued quasars we find only two in the Her II field, z = 1.09 and 1.13. They are further out than the z = 2.4 quasars but still very close. *Just glancing at the Her II field shows unmistakably this nest of quasars and high redshift objects closely grouped around the disrupted blue radio galaxy.* By this time there should be no need to compute probabilities. It is only required to notice that these objects are associated together in a group—just like the many previous cases demonstrated over the years.

There is more, however, to be gleaned from these hard won observational projects. In this case there is another radio survey field overlapping the field just discussed. As Figure E-2 shows the Her I field contains four blue radio galaxies. But the one at the top is very much like a parent galaxy of z = .02 with a quasar-like companion of z = .29. The blue radio galaxy at the left of the field, however, is only about 8 arc minutes away from a Westerbork radio quasar of z = .546.

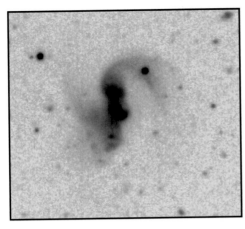

Fig. E-3. The blue radio galaxy at the center of the group of quasars in the upper right of Fig. E-2. Image courtesy William Keel.

But now comes a rather spectacular climax. Figure E-4 shows a blow up of the Her II field with all catalogued quasars plotted. There are three of high redshift (characteristically not radio sources). One falls close to the blue radio galaxy at the top of the field - a typical close companion. The other two fall across the brighter z = .55 radio quasar. Now compare this configuration with the two Arp/Hazard triplets shown in Figure 8-14 in the main text! The redshifts of the central quasars as well as the high redshift ejected quasars are almost identical. *The new triplet is, like the previous ones, aligned almost as exactly as the points can be plotted.*

The irony of this last piece of evidence is that I can remember vividly the astronomer who measured these high redshift quasars at a meeting in Santa Cruz many years ago. He stood up after a talk by Geoff Burbidge and said roughly: "The reason everyone rejects the association of quasars with nearby galaxies is that it has never led to any useful progress." After about 15 years my response seems to be: "The reason we have not had any useful progress is that astronomers don't even look at their own observations. "

The Quasar 2.4 Arc Seconds from a Dwarf Galaxy

As this book was going to press an announcement appeared in *Astronomy and Astrophysics Letters* that a QSO of z = .807 had been found to be a "by chance projection" on the center of a galaxy of z = .009. The chance by accident worked out to be about one in a thousand even if they had looked at every possible galaxy in their survey. Of course this also ignored all of the many other close juxtapositions of low redshift galaxies and high redshift quasars found previously.

But I worried that the dwarf, though slightly unsymmetrical, did not look like the kind of galaxy usually responsible for ejection of a quasar. So I looked around this pair. And what did I find? Nothing but a hugely bright (V = 10.98 mag.) Seyfert galaxy only 37 arc minutes distant (Figure E-5). This is just in the distance range where we found quasars to be systematically associated with Seyferts in Chaps. 1 and 2.

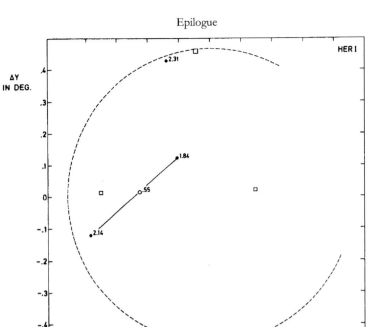

Fig. E-4. The Hercules I field enlarged with all catalogued quasars plotted. Note how the high redshift quasars are close to the blue radio galaxies. Particularly note the z = 1.84 and 2.14 pair aligned exactly across the z = .55 radio quasar. For comparison to previous Arp/Hazard triplets see Fig. 8-14.

The redshift of this Seyfert was z = .008 making a very strong case for the dwarf at z = .009 to be entrained material ejected from the Seyfert along with the quasar. Therefore the quasar did not have to be ejected from the dwarf. And, as a matter of fact, there is evidence of dwarf galaxies being ejected out in lines from active galaxies in deep photographs of NGC4651 (see *Astronomy & Astrophysics* 316, p63 Figure 6). Actually the dwarf did not even have to be ejected out with the quasar. It was shown in Chap.3, Figure 3-27, that quasars and companions are ejected preferentially along the same minor axis direction. The dwarf near the z = .807 quasar could have been entrained earlier. (The quasar shows no reddening, as it should if it were behind the dwarf. The pair should be studied with high resolution imaging and spectra to see if interaction can be detected.)

There are times of great tension in research. Now it was: "Would there be another quasar out along this line of ejection to the z = .807 quasar?" The answer was: Yes there was - and what's more it was of lower redshift! Then, look on the other side! There were two quasars much closer in along the line of ejection. They were of much higher redshift and, as quasars near z = 2 should be, about two magnitudes fainter than the lower redshift quasars.

This is all shown in Figure E-5. In that figure there is a line drawn through the Seyfert. The viewer would think that the line had been drawn through the line of the

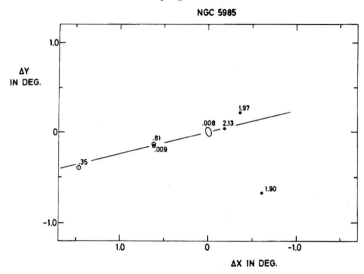

Fig. E-5. All quasars and bright Seyferts in the field of the z = .807 quasar and the z = .009 dwarf galaxy are plotted. The Seyfert is very bright at V = 11 mag. and the quasars decrease in redshift with distance from it. The line drawn in is the catalogued direction of the Seyfert's minor axis.

quasars. But that is not true. What I did was go to the Nilson Catalog of galaxies and look up the position of the minor axis of the central Seyfert, NGC5985. Before I looked at the number for the minor axis position angle I paused and thought, "It would be so conclusive if the minor axis came out along the line of quasars." My next thought, with considerable anguish was, "The chances of that happening are so small I have to be prepared to be disappointed." *But then I saw the minor axis come out along the line of the quasars as well as I could have drawn it!*

So this is a fitting end to the book. A confirmation of the sum of all the 32 years of observational evidence in Figure 9-3, the confirmation with the single Seyfert, NGC3516, association shown in Figure 9-7 and now finally the best aligned group of all with NGC5985.

One quantitative note: Figure E-6 shows the relation between the redshift of the quasar and its distance from the ejecting galaxy. The fact that it varies as the logarithm of the distance means the projected distance—redshift law is exponential. The fact that the slopes of the line are the same for both NGC3516 and NGC5985 means the law is the same for both systems. The displacement in ln r between the two relations means the scale of the NGC5985 distances is about 4.5 times the scale of the NGC3516 distances. Is this reasonable? First, the dereddened apparent magnitude of NGC5985 is 1.33 magnitudes brighter. If the two galaxies are the same luminosity, this implies NGC5985 is about a factor of two closer. The remainder of the factor is partially taken up by the fact that the listed inclination to the line of sight is greater for NGC5985. So

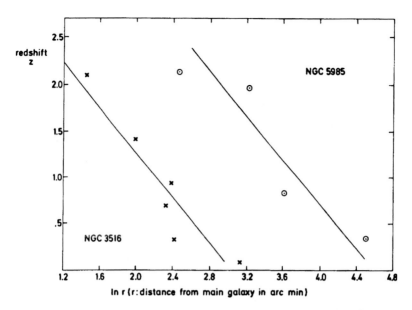

Fig. E-6. The relation between redshift and distance of the quasars along the
ejection lines from the two Seyferts is shown. The observed properties of
NGC5985 indicate it is closer to the observer than NGC3516 with its minor
axis more along the line of sight thus accounting for the wider apparent
spacing of the quasars.

the deprojected, absolute spacings between the quasars are quite similar for the two
cases.[2]

Finally there is an irony involved in this last discovery which in many ways
encapsulates the entire 32 year history of the association of high redshift quasars with
low redshift galaxies. The same principle author on the paper which stated so
authoritatively that the quasar 2.4 arc seconds from the dwarf galaxy was a "chance
projection" of a background quasar, had signed himself as a referee on the paper
concerning NGC1097 discussed in Chap. 2. That paper, reporting new X-ray
observations with three kinds of detectors on a Seyfert galaxy which had been shown
previously to have about thirty quasars associated with it, had been rejected without
chance of rebuttal. This same referee has now failed to look in the close vicinity of his
"chance projection" and therefore failed to make a clinching confirmation of the
ejection of quasars from active galaxies.

Not recognizing key observational evidence such as that around NGC5985 is
partially a consequence of suppressing observations like those in the NGC1097 paper
referred to earlier. All this has led to public statements such as, "Early claims of
disproof of redshift-distance assumptions have not been subsequently supported."
Unfortunately it is clear by now that this will certainly continue until enough people
start *Seeing Red.*

---

[2]   A quasar of z = .69, originally overlooked, falls 48.2 arcmin NE of NGC5985 and within 15 degrees of
the line in Figure E-5. In Figure E-6 it furnishes another point very closely on the angular separation-
redshift relation.

Glossary

Absolute magnitude
The brightness (measured in magnitudes) that an object would have if observed from a distance of 10 parsecs (32.6 light years).

Absorption line
Energy missing from the spectrum of an object in a narrow range of wavelengths, owing to absorption by the atoms of a particular element. The spectrum shows a black line where a characteristic color line would appear in case of emission of the same wavelength by the atoms.

Active galaxy
A galaxy with extremely high emission of radiation especially in the high-energy range: UV-radiation, X-rays, Gamma rays. Well-known examples are Seyfert galaxies, Markarian galaxies, radio galaxies, BL Lac objects, and quasars.

A *posteriori* probability
The probability, after an event has occurred, that it would occur.

Apparent magnitude
The brightness that an object appears to have at its actual distance, measured in magnitudes. (The faintest stars visible to the unaided eye are about 6th magnitude, and the faintest stars and galaxies photographed in large telescopes are about 30th magnitude).

A *priori* probability
The probability, before an event has occurred, that it will occur.

Barred spiral
A spiral galaxy in which the spiral arms unwind from a spindle shaped "bar" of stars that forms the galaxy's inner region.

Big Bang theory
The theory that the universe began its expansion at a particular point in space-time.

Black hole
A singular region in space within which gravitational force is so intense that no matter or light can escape.

BL Lac objects
Objects with spectra dominated by non-thermal, continuum radiation. Morphologically a transition between quasars and galaxies. Marked by very strong radio and X-ray emission.

Blueshift   The fractional amount by which the features in the spectrum of an astronomical object are shifted to shorter (bluer) wavelengths.

Boson   An elementary particle whose spin quantum number is an integer. Bosons are responsible for the effects of the forces of nature. An example is the photon which is the origin of the electromagnetic force.

BSO   Blue stellar (appearing) object.

B stars   Hot, luminous stars generally in an early stage of stellar evolution.

Butcher-Oemler effect   Surprising result that galaxies in higher redshift clusters tend to be bluer.

Bremsstrahlung   Radiation emitted by a charged particle which is de-accelerated if it encounters an atom, molecule, ion *etc.*

CCD   "Charge coupled device": Light-sensitive electronic chips used in modern astronomy to record and to measure the light received.

Celestial poles   The points on the sky directly above the Earth's north and south poles.

Chain of galaxies   A group of four or more galaxies that roughly form a line on the sky.

Compact source   A region emitting large amounts of visible, radio or X-ray energy from a small apparent area on the sky.

Companion galaxies Smaller galaxies accompanying a large, dominant galaxy in a galaxy pair or group.

cz   Redshift expressed in units of the speed of light ($c = 300,000$ km/sec)

Dark matter   Matter invisible to present astronomical instruments.

Deconvolution   A mathematical operation which helps to restore the true characteristics of an observed object. If the influence of the instrument (*e.g.* the point spread function) is known, the process allows the actual shape and intensity of the object to be better seen.

Declination	An angular positional coordinate of astronomical objects, varying from 0 degrees at the celestial equator to 90 degrees at the celestial poles.
Discordant redshifts	Redshifts which are other than expected at the distance of the object.
Δz	The difference between two redshifts: $z_1 - z_2 = \Delta z$.
E galaxy	A galaxy with smooth, ellipsoidal spatial distribution of predominantly older stars.
Electromagnetic radiation	Streams of photons that carry energy from a source of radiation.
Electron	An elementary charged particle, a constituent of all atoms, with one unit of negative electric charge.
Electroweak force	The unification of the electromagnetic force and the weak force. (An example for the latter is the decay of the neutron into a proton and an electron.)
Emission line	A "spike" of excess energy within a narrow wavelength range of a spectrum, typically the result of emission of photons from a particular type of atom in an excited state.
Excited state	An orbital state of an atom in which at least one electron occupies an orbit larger than the smallest allowed orbits. If an electron jumps to a lower orbit it emits a photon with an energy characteristic of the separation of both orbits. The result is an emission line in the spectrum of the atom.
Experimentum crucis	A decisive experiment that will prove or disprove a theory.
Frequency of radiation	The number of times per second that the photons in a stream of photons oscillate, measured in units of hertz or cycles per second.
Galactic equator	The plane of our Milky Way galaxy projected on the sky.
Galactic rotation	The collective orbital motion of material in the plane of a spiral galaxy around the galactic center.

Galaxy

An aggregate of stars and other material which forms an apparently isolated unit in space, much larger than star clusters (which are normal constituents of galaxies).

Gamma rays

A particular type of electromagnetic radiation of very high frequency and very short wavelength. Its origin is processes within the nucleus of an atom.

Globular cluster

A star cluster of spherical shape containing up to several 100 000 stars of very high age. Globular clusters form a spherical halo around the Milky Way and other galaxies.

Grand Unification Theories

Theories that try to unify all forces in nature.

Gravitational lens

An object with a large mass that bends the paths of photons passing close to it.

H_0

The Hubble constant defined as the ratio of a galaxy's redshift to its distance (distance often estimated from its apparent magnitude); its value is generally quoted as $H_0 = 50$ to 100 km s^{-1} Mpc^{-1}.

H I

Neutral (non-ionized) hydrogen, usually observed by radio telescopes, which detect the radio emission arising from the transition between different states of spin alignment of the atom's electron and the proton in its nucleus.

H II region

A gaseous clump of predominantly ionized hydrogen, excited by young, hot stars within it, and which therefore shows conspicuous emission lines.

Hertzsprung Russell Diagram (HRD)

A diagram representing the evolution of absolute magnitude versus color (a measure of temperature). Each star is represented by one point in a HRD.

Host galaxy

A galaxy with an active object (*e.g.* a quasar) at its center.

HRI

High Resolution Instrument on the ROSAT X-ray Telescope

Hubble's Law

The proportionality between a galaxy's redshift and its apparent magnitude.

Hydrogen α line

An important spectral line originating in the hydrogen atom, often seen as hydrogen α line emission in H II regions.

Im galaxies	Irregular, usually Magellanic Cloud type galaxies
Image processing	An analysis of images which renders contrast differences, gradient changes, discontinuities, and other systematic characteristics visible; nowadays best performed by computer algorithms applied to digitized data.
Isophotes	Lines connecting points of equal intensity on a skymap.
Jet	A linear feature, much longer than it is wide, usually straight, and inferred to arise from collimated ejection of material.
Late type galaxies	Galaxies showing a rotational disk and increasing amounts of young star population.
Light year	The distance light travels in one year, approximately 6 trillion miles or 10 trillion kilometers.
Local Group	The small cluster of about 20 galaxies that includes our Milky Way and the giant spiral (Sb) galaxy, the Andromeda Nebula (M31).
Local Supercluster	The largest nearby aggregation of groups and smaller clusters of galaxies, with the rich Virgo Cluster of galaxies near its center.
Luminosity class	Classification scheme of stars according to their luminosity. It extends from class I for supergiants to class VI for white dwarfs.
M	"Messier." A catalogue of nebulae, clusters, and galaxies compiled by Ch. Messier in 1784 (*e.g.* M87).
Mach's principle	A postulate put forward by Ernst Mach which states that inertia (mass) is the result of the influence of all particles within the universe. This contradicts the view that mass is an attribute of each single particle.
Magnitude	A measure of objects' brightness in which an increase by one magnitude indicates a decrease in brightness by a factor of 2.512.
Maser	Stimulated emission of electromagnetic radiation (*e.g.* by water molecules) in the microwave range.
Metallicity	The relation of the abundances of heavy elements to the abundance of hydrogen within stars.

Milky Way Our own galaxy, a spiral galaxy in the Local Group of galaxies.

Minor axis The axis about which a galaxy rotates. (It is perpendicular to the disk).

Mpc (megaparsec) One million parsecs.

Narlikar-Das mechanism The slowing down of the velocity of newly created matter particles in order to conserve momentum as they gain mass with time.

NGC "*New General Catalogue* of Nebulae and Clusters of Stars." A catalogue published in 1888 by J. Dreyer. It contained 7840 star clusters, nebulae, and galaxies. Appendices (called IC = *Index Catalogue*) extended it to more than 13,000 objects.

Noncosmological redshift A redshift not caused by the expansion of the universe.

Nonvelocity redshift A redshift not caused by velocity of recession.

North Galactic Hemisphere The half of the sky, divided by the galactic equator, that includes the north celestial pole.

Objective prism A wedge-shaped glass that provides small spectra of an entire field of bright sources.

O stars The hottest stars (50,000 K or more).

Opacity A measure which quantifies the amount of non-transparency of a medium. It is dependent on density, temperature, and chemical composition of matter.

Parsec A unit of distance, equal to 3.26 light years.

Peculiar galaxy A galaxy which does not have the standard, symmetrical form of most galaxies.

Pencil Beam Survey A survey of extragalactic objects using a narrow opening angle and extending to the detection limits of the instrument. It is believed to give information on the large scale structure of the universe.

Photon The elementary particle that constitutes light waves and all other types of electromagnetic radiation.

Planck particle	A hypothetical elementary particle with a mass of 5 000 000 trillion times the mass of a hydrogen atom. The particle is unstable and decays immediately after its creation into subparticles which subsequently decay to the constituents of ordinary matter (quarks, electrons *etc.*).
Point Spread Function (PSF)	The mathematical function describing now the light of a point-like source is spread out while passing through an astronomical instrument. The resulting image is not a point but a small disk whose radius is determined by the interaction of the radiation with the instrument.
Probability of association	If no physical association exists between objects, the probability that an observed configuration is a chance occurrence. Technically it is one minus the chance probability.
Quantization	The property of existing only at certain, discrete values.
Quantum gravity	A theory which tries to unify general relativity and quantum theory.
Quasar	A pointlike source of light with a large redshift, often a source of radio and X-ray emission as well.
Quasi-steady state cosmology (QSSC)	The cosmological theory that the universe is infinite in space and time and expands forever. Periodic explosive matter creation events ("mini-bangs") lead to an oscillatory motion of space superposed on the general expansion.
Radio lobe	Radio emission from appreciably extended areas on either side of a galaxy, often connected to the galactic nucleus by a radio-emitting jet.
Radio source	An astronomical object that emits significant amounts of radio waves.
Redshift	The fractional amount by which features in the spectra of astronomical objects are shifted to longer (redder) wavelengths.
Redshift-distance law	The hypothesis that an object's distance from us is proportional to its redshift (the usual interpretation of Hubble's law).
Redshift periodicity	The tendency of observed redshifts to occur with certain values at certain well-defined intervals from one another.

ROSAT   The German built, X-ray (*Röntgen*) Telescope

Right ascension   An angular coordinate of an astronomical object, measured eastward around the celestial equator (0 to 24 Hours) from the vernal equinox.

Schmidt telescope   A telescope with both a reflecting mirror and a correcting plate which can photograph a relatively large portion on the sky without distortion.

Seyfert galaxy   A special type of active galaxy (mostly spirals) detected by C. Seyfert. Seyfert galaxies are characterized by extremely bright cores whose luminosity shows extensive variability. They are also bright in infrared radiation and X-rays.

Solar motion   The motion of the sun with respect to nearby galaxies, which includes the sun's rotation around the center of the Milky Way as well as its peculiar motion within our own galaxy.

South Galactic Hemisphere   The half of the sky, divided in two by the galactic equator, that includes the south celestial pole.

Spectrum   The intensity of light from an object at each wavelength observed, using a prism or a grating. The result is a sequence of colored lines or strips characteristic of the chemical elements that emit the light.

Spiral galaxy   A galaxy in which the bright stars and interstellar gas and dust are arranged in a rotating, flattened disk within which prominent spiral arms of young stars and H II regions are visible. Spiral galaxies are classified as Sa, Sb, Sc (or SBa, SBb, SBc... if they are barred spirals). This sequence represents decreasing diameters of the central bulge and increasing separation of the individual spiral arms. An "I" added to the classification supposedly indicates high luminosity.

Starburst galaxy   A galaxy with an exceptionally high rate of star formation.

Steady-state theory The theory that the universe, on large distance scales, remains forever the same.

Supergalactic coordinate system   A reference coordinate system for external galaxies. The Local Supercluster, whose center is in Virgo, is concentrated around supergalactic latitude zero.

Supernova

An exploding star, which becomes (temporarily) thousands of times more luminous than the brightest normal star in galaxy.

Synchroton radiation

Radiation emitted by charged particles moving at nearly the speed of light whose trajectories are bent in a magnetic field.

Tidal interactions

Interactions between stars or galaxies due to their mutual gravitational attraction.

Tully-Fisher relation

For rotating galaxies, a correlation found by R.B. Tully< and J. R. Fisher between the luminosity and the width of the 21-cm radio line. It allows, in principle, to estimate the mass and hence the luminosity of a galaxy from the profile of its 21-cm line.

Universe

All observable or potentially observable matter that exists.

Virgo Cluster

The nearest rich cluster of galaxies, centered in the constellation Virgo.

Wavelength

The distance between two successive wave crests in a series of sinusoidal oscillations.

White Hole

A singular region in space-time, the time-reversed analog of a black hole, from which matter "falls out."

X-rays

A particular type of electromagnetic radiation, of high frequency and short wavelength.

X-ray source

An astronomical object that emits significant amounts of X-rays.

z

The symbol for redshift, defined as the displacement of spectral features in wavelength, expressed as a fraction of the original wavelength $z = \Delta\lambda/\lambda$.

z_0

Redshift corrected for solar motion.
Also used to denote the redshift of matter created at time t_0

List of Plates

Plate 1-7. From the ROSAT high resolution X-ray telescope. The strong X-ray Seyfert/Quasar Markarian 205 is shown ejecting X-ray filaments. On the ends of two of these filaments are quasars of much higher redshift. See Figure1-7.

Plate 2-7. Deep plates with the CTIO 4 meter telescope in red and blue were processed by Jean Lorre to yield this true color photograph of the jet Seyfert, NGC1097. Note apparent reddening of counterjets.

Plate 2-8. High resolution X-ray image of central regions of NGC1097—shown in false color with faintest surface brightness regions in red. Quasars nos. 26 and 27 are bright X-ray sources in direction of material leading from the active nucleus.

Plate 4-10. True color image of NGC7603 at $cz = 8,000$ km/sec attached by a luminous filament to a companion galaxy of $cz = 16,000$ km/sec. Photograph by Nigel Sharp and C.R. Lynds.

Plate 5-18. Contoured gamma ray counts in the Virgo Cluster for energies greater than 100 MeV to past 1000 MeV (where the sensitivity of EGRET becomes small). The quasar 3C279 is at a relatively weak phase where the connection to 3C273 is unmistakable. From a study by Hans-Dieter Radecke.

Plate 7-7. Hubble Space Telescope picture, in false color, of the Einstein Cross. At the wavelength of redshifted Lyman alpha there is connecting material between the right hand quasar and the central galaxy.

Plate 7-15. The Seyfert Galaxy NGC5252. The different colored arcs indicate redshift of the gas from +100 km/sec (red) to -100 km/sec (blue). Picture from J.A. Morse, J.C. Raymond and A.S. Wilson.

Plate 7-20. Photograph of a 4 square arc min field from 150 orbits of the Hubble Space Telescope. Note the dominance of peculiar and disturbed forms and the lack of normal appearing galaxies. Are they high or low luminosity? From Hubble Deep Field team (STScI) and NASA

Plate 8-18. M87, an active source of quasar and galaxy creation in the Virgo Cluster of galaxies as discussed in Chapter 5. This picture is in radio wavelengths by Frazier Owen.

Plate 8-19. This picture of Haro-Herbig 34 by Bo Reipurth shows a young star forming system in our own galaxy (HH34) in the combined light of sulfur and hydrogen emission. (From *ESO Messenger* No. 88, June 1997, p20). Note the resemblance to the jet in M87 in the preceding picture. Also the outer ejections show similarities.

Plate 8-20. The full field picture of HH34 in normal orientation (North at top, East at left). Note the outer ejection features show similarities to the outer M87 features.

Index

Elvis, Martin, 32
ESO Catalogue of Southern Galaxies
  ESO 161-IG24, 93
European Council, 102
European Southern Observatory, 26, 32,
  93, 125
European Southern Observatory Messenger, 146,
  215, 297

F

Fabian, A.C., 185
Fairall, Tony, 46, 161
Falomo, Renato, 240
Federation of American Scientists, 260, 262
Ferris, Timothy, 255
Festa, R., 220, 222
Field, George, 189, 190, 191, 236, 297
Fingers of God, 69
Finlay-Freundlich, E., 7, 98
Fisher, J.R., 295
Flandern, Tom Van, 268
Fleischmann, Martin, 264
Fort, B., 178
Fosbury, Robert, 26, 184
Foundation of Physics Letters, 255
Freud, Sigmund, 273
Friedmann, Alexander, iv, 225, 226, 248,
  249, 250
Frontiers of Fundamental Physics, 267

G

Gaia hypothesis, 269
Galaxies
  active, 287
  compact groups, 75, 76, 77
  companion, iii, 20, 29, 50, 53, 54, 57, 61,
    62, 63, 64, 65, 66, 71, 72, 73, 74, 75,
    76, 77, 78, 79, 82, 83, 84, 85, 86, 87,
    88, 89, 90, 95, 103, 104, 106, 107, 108,
    111, 112, 113, 120, 167, 179, 180, 186,
    188, 197, 202, 203, 208, 218, 240, 241,
    242, 243, 244, 245, 247, 297
  dominant, 26, 29, 64, 68, 72, 73, 75, 85,
    95, 288
  double, 62, 198
  merger, 14, 61, 72, 75, 104, 105
  protogalaxies, 72, 88, 118, 119, 163, 166,
    167, 223, 237
  spiral, 294
Galilei, Galileo, 259, 273
Gamow, George, 237
GC0248+430, 53
General Relativity, 225, 250, 254, 255

Giovanelli, Riccardo, 125, 150
Giraud, Edmond, 66
Goldsmith, Donald, i, ii
Goodstein, David, 270
Gravitational redshift. *See*
  Redshift:gravitational
Green, Richard, 98, 123
Gropius, 273
guilds, 274
Gunn, James, 200
Guthrie, Bruce, 198, 199, 200

H

h + Chi Persei, 99, 100
Hale Observatories, 27
Hale, George Ellery, 275
Halpern, J.P., 193
Haro-Herbig Catalogue
  HH34, 297
Harvard patrol camera, 132
Harvard University, 97, 132, 200
Hasinger, Günther, 42
Haynes, Martha, 125
Hazard, Cyril, 213
He, X.T., 43
Hemisphere
  Northern, 122, 292
  Southern, 95, 158, 163
Henry, J.P., 186, 189
Hercules, 280
Hermsen *et al.*, 132
Herrenstein and Murray, 271
Hewitt and Burbidge Catalogue, 204
Hickson, Paul, 75
High-energy Gamma Ray Observatory, 132
Hill, Steinhardt and Turner, 201
Holmberg, Erik, 84, 87
Holy Grail, 59
Hoyle, Sir Fred, 92, 108, 135, 137, 157, 173,
  229, 236, 237, 238, 244, 265
Hubble constant, 65, 68, 232, 234, 235, 249,
  250, 290
Hubble Deep Field, 190
Hubble diagram, 56, 235
Hubble relation, 3, 4, 21, 43, 54, 56, 61,
  68, 69, 80, 117, 154, 157, 182, 202, 232,
  236, 254, 290, 293
Hubble Space Telescope, 22, 23, 55, 64,
  145, 175, 189, 191, 247, 265, 280, 297
Hubble, E., 3, 4, 8, 22, 55, 64, 142, 148,
  153, 154, 189, 232, 235, 275, 290
Huchra, John, 68, 200
Humphreys, Roberta, 99
Hutchings, John, 55, 96

Markarian 205

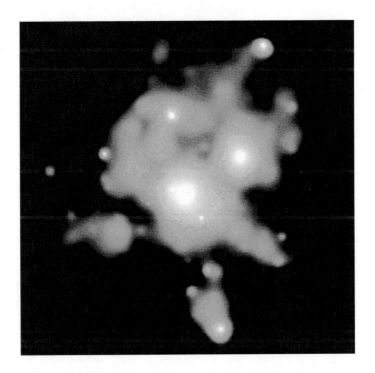

Plate 1-7

NGC 1097

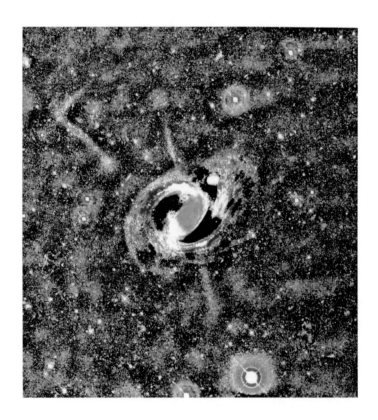

Plate 2-7

NGC 1097

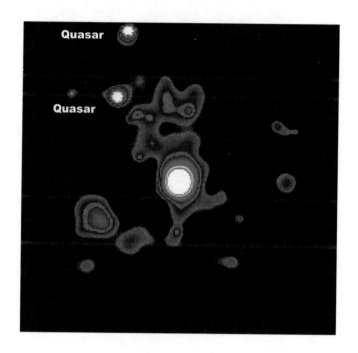

Plate 2-8

NGC 7603

Plate 4-10

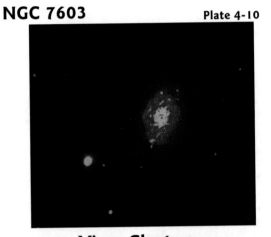

Virgo Cluster

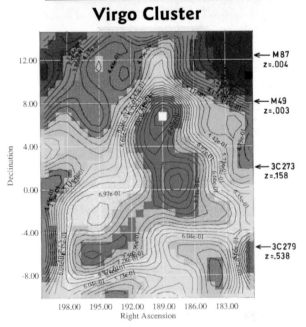

Plate 5-18

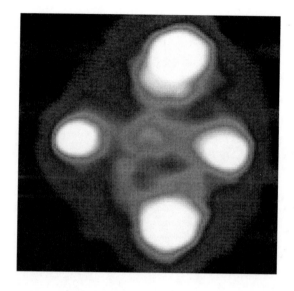

Einstein Cross

Plate 7-7

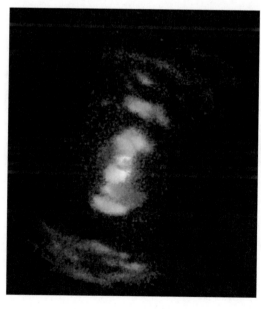

NGC 7252

Plate 7-15

Hubble Deep Field

Plate 7-20

M87

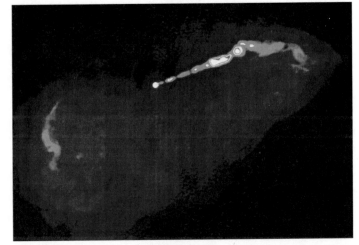

Plate 8-18

HH 34

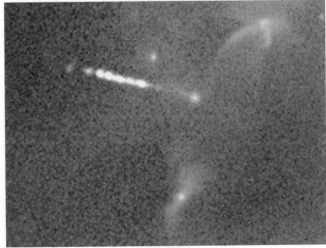

Plate 8-19

HH 34

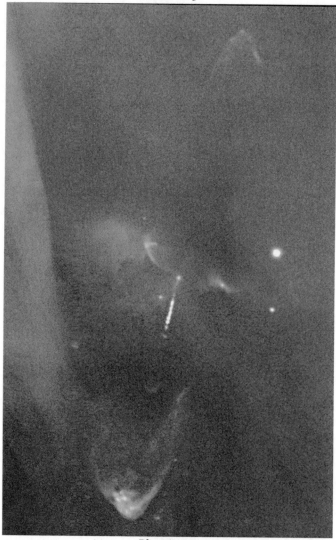

Plate 8-20

39543291R00179

Made in the USA
Middletown, DE
17 March 2019